Springer Theses

Recognizing Outstanding Ph.D. Research

Aims and Scope

The series "Springer Theses" brings together a selection of the very best Ph.D. theses from around the world and across the physical sciences. Nominated and endorsed by two recognized specialists, each published volume has been selected for its scientific excellence and the high impact of its contents for the pertinent field of research. For greater accessibility to non-specialists, the published versions include an extended introduction, as well as a foreword by the student's supervisor explaining the special relevance of the work for the field. As a whole, the series will provide a valuable resource both for newcomers to the research fields described, and for other scientists seeking detailed background information on special questions. Finally, it provides an accredited documentation of the valuable contributions made by today's younger generation of scientists.

Theses are accepted into the series by invited nomination only and must fulfill all of the following criteria

- They must be written in good English.
- The topic should fall within the confines of Chemistry, Physics, Earth Sciences, Engineering and related interdisciplinary fields such as Materials, Nanoscience, Chemical Engineering, Complex Systems and Biophysics.
- The work reported in the thesis must represent a significant scientific advance.
- If the thesis includes previously published material, permission to reproduce this must be gained from the respective copyright holder.
- They must have been examined and passed during the 12 months prior to nomination.
- Each thesis should include a foreword by the supervisor outlining the significance of its content.
- The theses should have a clearly defined structure including an introduction accessible to scientists not expert in that particular field.

Indexed by zbMATH.

More information about this series at http://www.springer.com/series/8790

Daniel Montero Álvarez

Near Infrared Detectors Based on Silicon Supersaturated with Transition Metals

Doctoral Thesis accepted by Universidad Complutense de Madrid, Madrid, Spain

Author
Daniel Montero Álvarez
Facultad de Ciencias Físicas
Universidad Complutense de Madrid
Madrid, Spain

Supervisor
Dr. Javier Olea Ariza
Facultad de Ciencias Físicas
Universidad Complutense de Madrid
Madrid, Spain

ISSN 2190-5053 ISSN 2190-5061 (electronic)
Springer Theses
ISBN 978-3-030-63828-3 ISBN 978-3-030-63826-9 (eBook)
https://doi.org/10.1007/978-3-030-63826-9

This Springer imprint is published by the registered company Springer Nature Switzerland AG
The registered company address is: Gewerbestrasse 11, 6330 Cham, Switzerland

Supervisor's Foreword

The world of infrared detection is certainly rich. There are many technologies and its development is recently speeding up because of the advent of new needs and commercial niches. Infrared detectors are now installed in cell phones, cars and virtual reality headsets among others, and are one of the essential devices for military purposes. Moreover, the set up of 5G and the Internet of Things will increase the already high infrared market. So, why it is still a relatively expensive technology? While a regular 12 megapixels cell phone camera could cost around 10 euros (probably less), an infrared camera with much less resolution would cost at least 30 times more (with no more than 640×480 pixels). Obviously, economies of scale will reduce this price in the near future, but still, it is not the main technological problem.

Infrared detectors, and now I am more concerned about focal plane arrays, are usually based on rare, expensive and pollutant materials (Pb, Ga, As, Hg, Cd, etc.), what makes the device much costlier. Also, bolometer detectors have the disadvantage of the speed in detection and a low sensitivity (detectivity). Finally, the use of these "rare" materials hinders the integration of the detector with the read-out integrated circuitry, which is absolutely necessary, and is usually based on silicon. This incompatibility is also a source of price increasing. Since most of the microelectronic technology is based on (the cheap) silicon, it would be desirable to have also an infrared detector technology based on silicon, working at room temperature, and with the possibility of being integrated into high resolution focal plane arrays, sensible, at least, up to the first couple of microns of the infrared range. Several of the described applications rely on this part of the infrared.

Since bare silicon does not detect the near infrared, one could think about the possibility of modifying and enhancing it. Now comes into play the intermediate band materials theory, which was firstly created for photovoltaics: if we could break the silicon bandgap into two different smaller bandgaps by supersaturating it with deep level impurities, and if the resulting material could have the adequate electronic transport and optical properties, we could have the Holy Grail of the infrared detection.

This book covers most of the research developed by Daniel about Ti supersaturated Si in the Thin Films and Microelectronics Group of the Universidad Complutense de Madrid in the last 5 years. Structural, optical and electrical analysis were conducted and the main conclusions are included. Also, the first integration efforts of Ti supersaturated Si layers into a Focal Plane Array CMOS Image Sensor are described within the frame of a collaboration with ST Microelectronics. Maybe we will have the opportunity to see the invisible light with this technology in a few years.

June 2020

Dr. Javier Olea Ariza
Facultad de Ciencias Físicas
Universidad Complutense de Madrid
Madrid, Spain

Abstract

Along this thesis, titled "Near Infrared detectors based on silicon supersaturated with transition metals", we describe the research based on Ti supersaturated Si substrates, aiming to extend the photoresponse of bare Si towards photon energies lower than the bandgap at room temperature. The starting material is a crystalline Si substrate, which is Ti ion implanted in concentrations up to five orders of magnitude higher than the solid solubility limit (hence the "supersaturation" term). Later, a Nanosecond Laser Annealing (NLA) treatment is used to recover the crystal quality, lost after the implantation process. When the concentration of Ti atoms is high enough, the discrete wave function of each impurity may overlap to form an allowed band of states, between the valence and the conduction band, called the impurity band. Thus, carriers from the valence band could promote to the conduction band through the Ti impurity band by absorbing photons with energy lower than the bandgap.

The purpose of the thesis is to develop a fabrication route that could integrate the Ti supersaturated Si material into a commercial CMOS Image Sensor, targeting the imaging in the Near Infrared (NIR) and Short-Wave Infrared (SWIR) ranges, that is, from 0.7 to 3.0 μm, at room temperature. There are several fields (outdoors night vision, smart driving, food inspection, disaster management, oil spill detection, Internet of Things) that are demanding cheaper IR imaging sensors with higher resolution. A material based on silicon, compatible with a CMOS route, could offer more competitive alternatives to what is actually available in the market. Most imaging solutions used in the NIR and SWIR range nowadays lack one or more of the next features: CMOS route compatibility (which has a direct impact on the cost), room temperature operation, fast response, high scalability (to provide resolutions, at least, of several megapixels), environmentally friendly and abundant raw materials. To date, uncooled Ge and InGaAs photodetectors are mostly used to detect photons in the SWIR range, which feature considerably higher prices per pixel than Si technologies.

In this thesis, we orient our research towards the industrialisation of the Ti supersaturated Si material, to which we designed and fabricated the first prototypes, in the microscale, in our facilities at UCM. The first set of microscale photodiodes showed promising results with responsivity, at room temperature, down to 0.45 eV (2.75 μm), while bare Si is usually responsive down to 1.12 eV (1.1 μm). We observed that the polarisation in reverse could increase the Quantum Efficiency (QE) up to two orders of magnitude. This first batch of devices, which still require further optimisation, showed that the Ti supersaturated material could be scaled down to the microscale, with potential utility as room temperature operated photodiodes with applications in the NIR and SWIR range.

The next step was to search for an industrial partner with which further develop the technology of supersaturated Si materials. We signed a collaboration with STMicroelectronics (Crolles, France) to perform a first integration of the material into a CMOS Image Sensor route, materialised as a six-month internship in their facilities. Before the internship, it was necessary to find appropriate suppliers that could provide certified equipment compatible with 300 mm wafers, for both the Ti implantation and NLA processes. We found two French suppliers, Ion Beam Services for the Ti implantation and SCREEN-LASSE for the NLA process. The laser featured a XeCl source with long pulses with duration of 150 ns.

Prior to the integration on the CMOS Image Sensor route, it was necessary to fabricate new Ti implanted samples annealed using the XeCl laser, aiming the find the best conditions that could lead to higher efficiencies. We studied the material properties and the photoresponse of macroscale photodiodes as a function of the Ti dose and the laser fluence. This preliminary study was key to understand the recrystallisation processes of Ti implanted samples with the XeCl laser and led to the definition of the fabrication parameters (Ti dose and laser fluence) that were used in the internship.

During the internship, we studied the different possibilities of integration of the Ti supersaturated Si material into different pixel structures. Finally, we chose a Back-Side Illuminated pixel configuration. We developed the integration plan, which was followed to fabricate several 300 mm wafers for material and device characterisation. The monitor wafers (used for material characterisation) showed that the Ti concentration levels were high enough to possibly form an impurity band in Si. We also determined which laser processes produced better crystalline layers. The knowledge acquired from the monitor wafers was useful to determine the best fabrication parameters of the final pixel matrices. The Ti supersaturated pixels showed slightly lower QE results than non-implanted pixel matrices for wavelengths longer than 500 nm (from green to NIR) up to 1000 nm, the limit of the equipment available at STMicroelectronics. We expect to measure the samples in the SWIR, up to 1.6 μm, in a new optical bench that is being installed at CEA-LETI (Grenoble) in the near future.

The integration of the Ti supersaturated material was successful: no Ti cross-contamination was found on neither the wafers nor the instruments used during handling and fabrication, which is key when analysing the possible industrialisation of a new material (Ti atoms are deep centres in low concentrations, which affect the rest of the technologies developed in the same fab in case of cross-contamination). Besides, the NLA process proved to be compatible with the pixel structure for laser fluences lower than a certain threshold. Finally, the Ti supersaturated pixel matrices showed homogeneity levels between pixels very close to the reference, which further evidences that the Ti supersaturated Si material is fully compatible with a commercial CMOS Image Sensor route.

The compatibility of the Ti supersaturated Si layers into a CMOS Image Sensor route is the first step towards the possible application of this material in commercial devices, sensitive in the NIR and SWIR range. The collaboration with STMicroelectronics is ongoing and is expected to continue with the optimisation of the Ti supersaturated Si material.

Acknowledgements

My passion for science has been part of me since I was a child. I guess my curiosity towards nature comes from my mother, María Isabel, while my passion for technology comes from my father, José María. When I was a child, I could spend countless hours looking at the night sky, wondering about the universe and all the mysteries yet to be resolved. My passion for science continued growing since then. When I was 16 years old I decided to do a Ph.D. in Astrophysics, aiming to discover as much as I could from the universe.

Several things have changed since then. My passion for Astrophysics was slowly replaced by my passion for renewable energies, semiconductors and electronics. Most of the fault to this change must be acknowledged to Ignacio Mártil, who taught me about the wonders of semiconductors in the fourth year of the degree. It only took two or three classes to go ask him how to start doing research.

There are so many people to thank along this journey. First, to my parents and sister. It would not have been possible without their support, help and understanding. Second, to the love of my life, Laura, because she has seen me at my best and also my worst. In those moments when you want to give up, she put me back in track, so I could continue to where I am now. I am sure that all the effort was worth it. I want to thank also all my friends who have dealt with me since the beginning, and those who came later. Special mention to Fernando, Elena, Sara and Adrián. We have spent countless hours studying hard and I am sure I owe them more than I can remember.

My journey towards the Ph.D. has led me to know so many people, from the Faculty and from the research group of "Thin Films and Microelectronics". Thanks to Javier Olea, for all the teaching, laughter and patience, he deserves a lot of credit. I miss the days when we travelled by car to Gant to the Photonics Summer School. To German González, because he has been like my godfather in the research area, especially when it comes to design/repair/modify/improve any aspect related to scientific equipment and scientific curiosity. His patience is also beyond limits and that is greatly appreciated. To Ignacio Mártil, as I mentioned before, as he introduced me to the scientific world. To Marisa Lucía Mulas, she is always nice and makes you feel welcome in the research group. To Álvaro del Prado, as he is always willing to help anybody and that is rare to find nowadays. To Enrique San Andrés, I learned a lot from him regarding microelectronics technology; he pushed my academic orientation towards the microelectronics, unknown for me before meeting him. To David Pastor, for all the helpful talk about lasers and van life. To Eric García, it is always nice to learn about semiconductors from him. To Rodrigo García, he is always in a good mood and we have had the best laughter when going to conferences. Thanks to Daniel Caudevilla and Sari Algaidy, it has been short but sure fun. Special thanks also to Pablo Fernández and Rosa Cimas, from CAI Técnicas Físicas for all their technical and scientific support. Thanks also to Antonio Paz and Fernando Herrera for their technical support and advice. If it were not for them, several instruments, from the sputtering system to the Reactive Ion Etching, would not have worked.

I also have to thank many people from France, where I spent six wonderful months of my life doing my internship in STMicroelectronics. I felt very welcome from the first day. Although not from France, thanks to Rodrigo Blasco, a classmate from the Master's degree, who introduced some very good friends in Grenoble: Akhil, Marion, Nathaniel, Mada, Saptarshi, with whom I have shared great moments. To my neighbours David Bastian and Clarie Paris, because they saved me from really awful moments up there in France. They welcomed me as a friend when I could not even speak French. At STMicroelectronics, to all the people that made possible the project, in special to François Roy and Olivier Noblanc, for believing in the project from the first day. The huge time dedication and effort of my workmates must be acknowledged as well: Andrej Suler, Boris Rodrigues, Sonnarith Chhun, Thomas Dalleau, Romuald, Bastien, Carlos, Nils, Yolenne, and so many people from the fabrication and characterisation processes, which pushed the wafers into the fab to deliver them on time. Finally, from CEA-LETI (Grenoble), special thanks to Sébastien Kerdiles and Pablo Acosta Alba. They were always helpful and I could learn most of my knowledge regarding the recrystallisation processes and laser interaction with matter, opening a new and fascinating world to me.

Daniel Montero Álvarez

Contents

Acronyms

AC	Alternate Current
ADC	Analog-to-Digital Converter
AFM	Atomic Force Microscopy
ARC	Anti-Reflecting Coating (layer)
BEOL	Back-End Of Line
BGN	Band-Gap Narrowing
BOE	Buffer Oxide Etchant
BSF	Back-Surface Field
BSI	Back-Side Illumination
CAI	Centro de Asistencia a la Investigación
CB	Conduction Band
CBD	Cellular Breakdown
CCD	Charged-Couple Device
CDTI	Capacitive Deep Trench Isolation
CIS	CMOS Image Sensor
CMOS	Complementary Metal-Oxide-Semiconductor
CMP	Chemical-Mechanical Polishing
CVF	Conversion Voltage Factor
DC	Direct Current
DMSO	DiMethyl SulfOxide
D-SIMS	Dynamic-Secondary Ions Mass Spectroscopy
DSNU	Dark Signal Non-Uniformity
DSP	Double-Side Polished
ECR-CVD	Electron-Cyclotron Resonance Chemical Vapour Deposition
EDX	Energy Dispersive X-ray Spectroscopy
EQE	External Quantum Efficiency
FEOL	Front-End Of Line
FFT	Fast Fourier Transform
FGA	Forming Gas Annealing
FIB-SEM	Focused Ion Beam Scanning Electron Microscopy
FIR	Far Infrared
FLA	Flash Lamp Annealing

FPA	Focal Plane Array
FPN	Fixed Pattern Noise
FSI	Front-Side Illumination
FWC	Full-Well Capacity
FWHM	Full Width at Half-Maximum
HM	Haze Measurements
HRTEM	High Resolution Transmission Electron Microscopy
IB	Impurity Band
IBS	Ion Beam Services
IPA	IsoPropyl Alcohol
IV	Current-voltage (curve)
LWIR	Long Wave Infrared
MIR	Mid-Wave Infrared
MOS	Metal-Oxide-Semiconductor
NDA	Non-Disclosure Agreement
NIR	Near Infrared
NLA	Nanosecond Laser Annealing
ONO	Oxide-Nitride-Oxide
PAI	Pre-Amorphisation Implantation
PCC	Photon Characteristic Curve
PD	Photodiode
PECVD	Plasma-Enhanced Chemical Vapour Deposition
PLM	Pulsed Laser Melting
PR	Photoresist
PRNU	Photon Response Non-Uniformity
PTC	Photon Transfer Curve
QE	Quantum Efficiency
R&D	Research and Development
RF	Radio Frequency
RIE	Reactive Ion Etching
RMS	Root Mean Square
ROIC	Read-Out Integrated Circuitry
RTA	Rapid Thermal Annealing
SEM	Scanning Electron Microscopy
SMU	Source and Measure Unit
SN	Sensing Node
SNR	Signal-to-Noise Ratio
SRH	Shockley-Read-Hall (recombination)
SRIM	Stopping Range of Ions of Matter (software)
SSP	Single-Side Polished
STEM	Scanning Transmission Electron Microscopy
SWIR	Short-Wave Infrared
TCAD	Technology Computer-Aided Design
TEM	Transmission Electron Microscopy
TG	Transfer Gate (transistor)

TIL	Titanium Implanted Layer
TLM	Transfer Line Method
ToF-SIMS	Time of Flight-Secondary Ions Mass Spectroscopy
T-R	Transmittance-Reflectance
TRR	Time-Resolved Reflectometry
TSL	Titanium Supersaturated Layer
TXRF	Total Reflection X-ray Fluorescence
UCM	Universidad Complutense de Madrid
UV	Ultraviolet
VB	Valence Band
VLWIR	Very Large Wave Infrared
V-PPD	Vertically Pinned-Photodiode
WLC	Wafer Level Characterisation
XTEM	Cross-section Transmission Electron Microscopy

List of Figures

List of Tables

Chapter 1
Introduction

1.1 The Role of Semiconductors

The vast majority of the people we know would not be able to adequately precise what are semiconductors used for, and even fewer people would be able to define what a semiconductor is. However, semiconductors have changed the modern society forever, as a subtle and silent revolution. Think about any particular day of our lives. From the very moment we wake up we already depend on semiconductors to start the journey, in the form of any electronic devices we could think of: from the alarm clock that rings in the morning, to the microwave, the TV, the cell phone, the computer, or any form of motorised transportation. Basically everything that surrounds us is made around electronic components that strongly depends on semiconductors to work. Semiconductors are revolutionising the society to levels we cannot imagine. And what is more impacting is that we have not fully exploited its true potential.

The first time we hear about semiconductors might have been when we studied the periodic table back in school. They are a rather small portion of the known elements (being metals the vast majority), yet they are extremely important. Effectively, as the word semiconductor suggests, they have properties that somewhat fall between those of a conductor (metals) and an insulator. In an extremely simplified vision on the underlying physics of semiconductors, their conduction properties strongly rely on the relative energetic position of the last occupied electron band (Valence Band, VB) with respect the first unoccupied band (Conduction Band, CB). This difference in energy is what is called the bandgap. Charge carriers are not allowed in this range of energies, so they must rely on external sources of energy (e.g. heat or electric field, among others) to pass from one to another. Band structure is strongly linked to the elements that form the semiconductor (typical examples could be Si, Ge as elements or GaAs, InSb or HgCdTe as compounds), as well as their periodic arrangement in the lattice (crystalline structure). Using this simplified vision, it is relatively easy to understand the electrical conduction properties of conductors: their VB and CB are very close to each other, or directly overlapping, so charge carriers

D. Montero Álvarez, *Near Infrared Detectors Based on Silicon Supersaturated with Transition Metals*, Springer Theses,
https://doi.org/10.1007/978-3-030-63826-9_1

(holes or electrons) can move freely without (much) resistance, thus showing high conductance. Insulators on the other hand have their VB and CB far in terms of energy, that is, a wide bandgap. A carrier must absorb a substantial amount of energy to be able to switch bands. Therefore, it is unlikely that these materials are electrically conductive unless their carriers absorb at least as much energy as the bandgap. The definition of a semiconductor comes out naturally using this (very) simplified version of the conductance model: their bandgap is halfway between those of conductors and insulators. A specific amount of energy is still needed to increase the conductance of the material, but it is in the order of a few electronvolts. There is a broad range of books and publications that widely covers the physics of semiconductors and that can be used for further reading [1–5].

The first experiments involving semiconductors date back to Alessandro Volta [6] in 1782. However, it was not until the 20th century that semiconductors started to gather the attention of the scientific community. The main kick-starter of the interest in semiconductors was the start of both World Wars as a necessity to improve radio emitters and receivers, as well as night vision sensors, mainly for military equipment. In the earlier years of the 20th century, PbS and Ge were mostly used to fabricate the first semiconductor devices, being the rectification and amplification of electrical signals the most valued characteristics. Silicon started to be considered as the semiconductor of choice around the 1950s, due to lower leakage currents than Ge [6]. The final boom in the use of semiconductors came after the invention of the integrated circuit by J. Kilby in 1958 and R. Noyce independently in 1959.

1.1.1 Silicon

Silicon is the seventh most abundant element in the universe, and the second on the Earth's crust, after oxygen [7]. It is more commonly found as part of silica (silicon dioxide), the most abundant component in sand. The abundance of silicon is one of the key features that makes it the undisputed king in the world of semiconductors, especially when applied to electronics and, by extension, microelectronics. Other characteristics that have led to the exponential development of silicon as compared to other semiconductors is, for example, the relative easiness to dope it, using also common elements as boron or phosphorus, mainly. Besides, the maximum solid solubility of said impurities is relatively high, in the order of 10^{20} cm^{-3}, which allows the doping of Si substrates in a wide range, increasing the number of applications [8]. Another key aspect of silicon is the passivating effect that its native oxide offers on the surface defects. This last property is key to develop planar devices (mainly transistors), where surface defects could severely compromise the reliability and uniformity of the integrated circuits [9]. Besides, the relative mechanical stability of silicon makes it suitable for easy handling during the fabrication processes. Finally, the refinement process of the raw material (sand or silica) into a monocrystalline structure, by means of the Czochralski method, is mature and reliable, obtaining perfect monocrystalline ingots of several meters long and up to 450 mm diameter with controlled doping [10]. The ingot is later cut into wafers, with almost perfect flatness, necessary to guarantee the best reproducibility and reliability of the fabricated devices. Thus, obtaining, refining and processing silicon is considerably cheaper than other semiconductors nowadays. It is estimated that silicon represents around 99% of the semiconductor market share in 2019, being GaN and SiC the next most used semiconductors after silicon [11, 12].

However, silicon features several drawbacks, the most important of them related to its electronic band structure. Silicon features an indirect bandgap, which results in a lower absorption coefficient than if it had a direct bandgap. The bandgap of silicon is 1.12 eV at room temperature, which makes it a good choice to absorb radiation in the visible. In fact, silicon is used in the vast majority of solar cells due to its relatively well-matched bandgap value with respect to the solar spectra, whose emission peak is located around 500 nm, in the visible [13]. Silicon is also the semiconductor of choice when it comes to photodetectors with applications in the visible region. The indirect bandgap of Si and its absorption coefficient imply that devices based on it should be around 150–200 μm thick to fully absorb the incoming light, especially radiation with lower energies, close to the bandgap. Thus, Si is not an efficient absorber of radiation when the energy is close to the bandgap; silicon is transparent to photon with energies lower than 1.12 eV, in the InfraRed (IR) part of the spectra.

1.2 The Infrared

The infrared is the part of the spectrum located at wavelengths longer than the red edge of the visible spectrum, from 700 nm, up to 1 mm, where the microwave region of the spectra starts. The discoverer of infrared radiation was W. Herschel [14] in 1800, when he measured an increase in the temperature of a thermometer when illuminated with light "beyond the red colour", invisible to the naked eye.

Several physical phenomena involve radiation in the IR. All matter whose temperature is above the absolute zero must emit radiation following a black body spectrum, as described by the Planck's law. Thus, most of the matter that surrounds us emits radiation, from the sun (approximately half of the energy delivered by the sun is in the form of IR radiation [15]) to everyday objects, people, animals, plants, everything. In the case of objects at room temperature, their peak emissivity is located at several microns of wavelength. Other physical phenomena, at the microscopic level, also emit radiation in the IR, for example the vibrational and rotational states of molecules and atoms, which is used in several branches of spectroscopy to identify the nature of the elements, including possible transitions and fundamental states of matter.

Several classifications were born to parcel the wide range of the electromagnetic spectrum covered by the infrared, depending on the field of application. In this thesis, we will follow the commonly accepted division for photodetectors, as suggested by J. L. Miller [16, 17]:

Near Infrared (NIR): from 0.7 to 1.0 μm. It is defined from the limit of detection of the human eye to the commonly accepted detection limit of Si at room temperature. Several applications use this band of the IR: in medicine at photodynamic therapy for cancer treatment, IR Raman spectroscopy in a wide variety of research areas, Time of Flight Sensors for smart and autonomous driving, cracks or failure detections in solar cells, IR reflectography in artworks to identify pigments and authenticity of the works, biochemistry and pharmacology or food inspection, among others.

Short-Wave Infrared (SWIR): from 1.0 to 3.0 μm. From the limit of detection of Si up to the first atmospheric absorption band of CO_2. It is the range of interest in this thesis. Ge and InGaAs detectors are sensitive in part of this range. The range of applications covers from fibre optic telecommunications, materials identification, wildfire response (SWIR light is not dispersed by the smoke), food security, geology, quantification and quality of the vegetation from satellites, oil spills or soil moisture detection [18]. Night vision in open environments due to the "night-glow" [19] light, even at moonless nights, is also possible using SWIR cameras.

Mid-Wave Infrared (MIR): from 3.0 to 5.0 μm. It extends up to the main atmospheric absorption band of H_2O, and it is covered by InSb and HgCdTe detectors mainly. The applications are mostly oriented through thermal night vision for surveillance and military uses.

Long-Wave Infrared (LWIR): from 5.0 to 12 μm. It covers the atmospheric window between the absorption band of H_2O and the fifth band of CO_2. The most used semiconductors are HgCdTe and microbolometers of different semiconductors, mainly vanadium oxides and amorphous silicon [20]. Most applications are related to the thermal emission of objects at room temperature. Thus, thermal night vision and material control inspection are highlighted. Thermal vision in this range is also used to detect thermal losses in edifications, anomalous heating of electronic components, solar cells, car engines or batteries.

Very-Long Wave Infrared (VLWIR): from 12 to 30 μm. Cooled Si detectors are used in this range or devices based on silicon photonic structures [21]. The applications are thermal night vision and research in the field of astronomy and astrophysics.

Far Infrared (FIR): from 30 μm to 1 mm. Technology in this range (frequencies in the range of THz) is not as developed as in the rest of ranges. Cooled gallium doped germanium is used for space applications [22]. The applications are mostly limited to space detection, astronomy and astrophysics.

According to this division, silicon would be responsive only in the NIR. Thus, applications that rely in the rest of the IR range must seek for technologies based on other semiconductors to fulfil their needs. As commented before, most of the semiconductor technology nowadays spins around silicon; the alternatives are considerably less mature and more expensive. Some of the alternatives are based on rare materials, as Ga compounds, are contaminant (like Pb salts) or operate at cryogenic temperatures [23]. The use of refrigerated devices seriously limits the range of applications; cooling systems are often heavy, bulky, expensive and need regular maintenance.

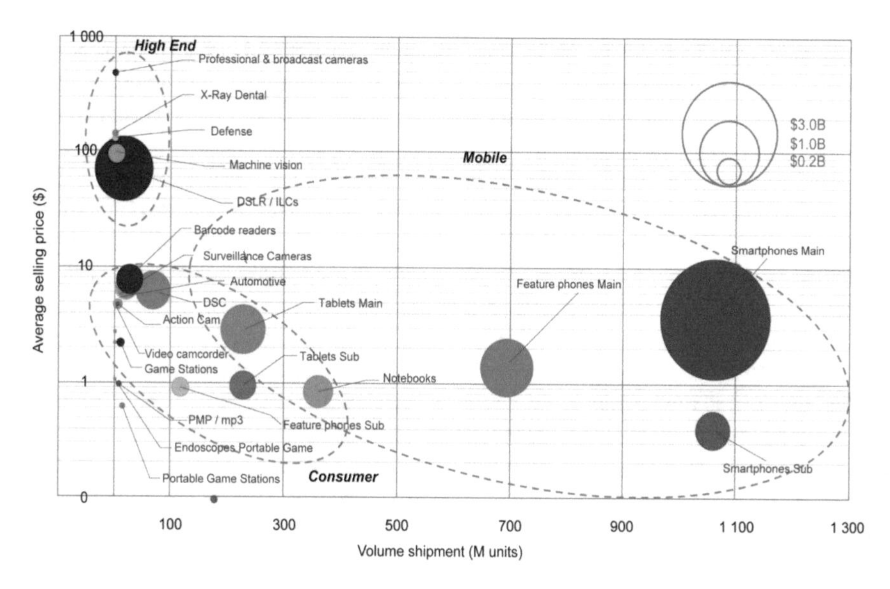

Fig. 1.1 Average selling price as a function of the shipment volume for several technologies used in the IR (*Source* "Infrared Detector Technology and Markets Trends Report 2015" from Yole Développement)

The number of applications that uses IR radiation is increasing each year, as the technology is slowly becoming more accessible to the big audience, due to the improvements that lower down the prices [24]. Is in this context where several efforts from both the research and the industry are focused in obtaining smaller, cheaper and more efficient NIR and SWIR detectors, especially in the form of Focal Plane Array (FPA) cameras [25]. Two main fields are pushing forward the research of new alternatives to sensing in the NIR and SWIR ranges: the smart driving for autonomous vehicles and the Internet of Things (Fig. 1.1).

Autonomous vehicles need a variety of sensors of different nature (what is called redundancy, to avoid failures) to adequately scan the surroundings: position, size, distance and speed of any object, in order to identify possible risks: pedestrians, other vehicles, animals, obstacles, traffic signals, bumps. In this context, there are several sensors that are of extreme interest in the automotive industry: LIDAR and Time of Flight sensors, used to determine the distance of any given object and night vision FPA cameras. The three possibilities rely on the use of NIR and SWIR radiation, produced by either laser of Light Emitting Diodes (LED) sources. Why NIR and SWIR radiation? Because it cannot be seen by humans nor most of the wildlife, and is less prone to be deviated by fog, smoke or haze. Some interesting interactive examples can be found elsewhere [26]. Therefore, light sources in said ranges can be used to illuminate the surroundings of the vehicle without perturbing the pedestrians, while also being more efficient under low visibility conditions [27]. Besides, the Internet of Things, being this term referred to the hyper connection of quotidian objects to the internet, is expected to increase the demand on NIR and SWIR sensors for similar reasons as in smart driving: distance sensors, detectors of presence or even wireless communication by means of IR light emitted from diodes (the so-called LiFi communication [28] under the Free Space Optical, FSO communication standard [29, 30]).

Another field that is demanding better devices for light sensing in the SWIR is photonics. Photonics, in contrast to electronics or microelectronics, uses light as the basic element carrying information, instead of electrons/holes. Photonics has experimented an exponential growth in the past years. A quick search in Scopus with the keyword "Photonics" reveals that from 2004 to 2014 the number of published works in peer-reviewed journals increased by a factor of ten. In particular, silicon photonics has attracted a lot of interest, especially in the field of fibre optic telecommunications. Since Si is transparent at the range of wavelengths used in this field (from 1.2 to 1.7 μm), several waveguide structures have been developed using bulk silicon, to benefit from the already existing knowledge inherited from the microelectronics industry. However, photonics still need cheap, efficient and easily integrated SWIR detectors, that is, materials absorbent in the same range of wavelengths in which silicon is transparent.

Figure 1.2 best represents the breakthrough a CMOS-compatible, IR Si-based camera could represent in the market. On average, CMOS Si sensors of smartphones feature resolutions in the order of 12–40 Mpx, and the cost is somewhere between 1 to 30 dollars per module. For example, the estimated price of the CMOS camera of the latest high-end smartphones is reported by TechInsights [31] and is located around 25 $. It would be of extreme interest if there could be a way to increase the absorption

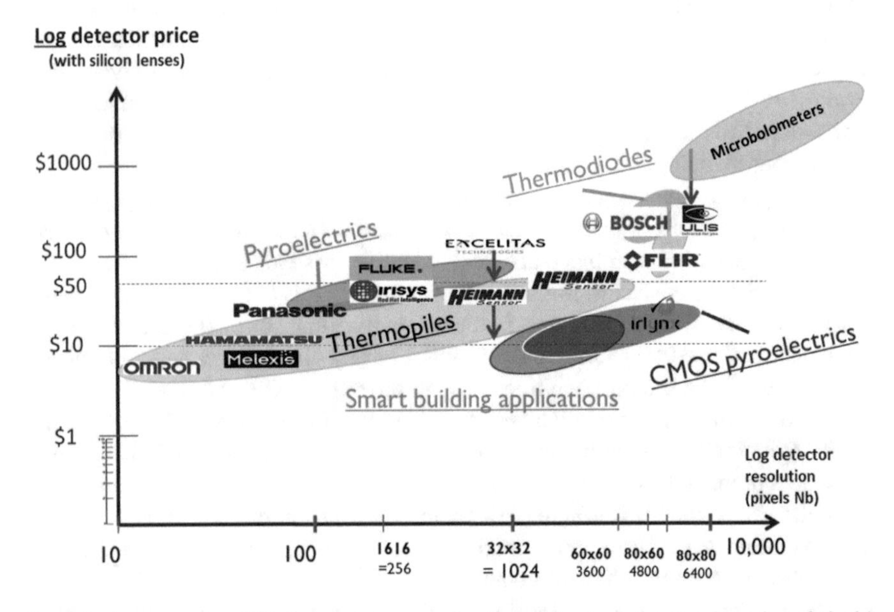

Fig. 1.2 Average price of IR photodetectors as a function of the resolution (total number of pixels) (*Source* "Infrared Detector Technology and Markets Trends Report 2015", Yole Développement)

coefficient of silicon so it could absorb photons with energies lower than its bandgap, that is, in the SWIR region. If such improvement could be accomplished, then most of the applications described in the previous lines could benefit from cheaper solutions, compatible with an already widespread and mature technology. A Si-based material that could operate at room temperature, with extended photoresponse in the SWIR range would be a game-changer in the future development of IR photodetectors.

1.3 Supersaturating Si

Several attempts have been done to increase the absorption of sub-bandgap photons on Si devices at room temperature. Some authors reported extending the Si responsivity up to 1.3 μm by using heavily doped Si to induce a Band-Gap Narrowing (BGN) effect [32–34]. The BGN effect uses the concept of the formation of an Impurity Band (IB) [35, 36]. The theory of impurity bands state that if the concentration of impurities is high enough, the discrete wave function of each impurity would overlap, turning the discrete levels into a continuous band of allowed states. In the case of shallow dopants, the IB would overlap with one of the bands at room temperature, producing the aforementioned BGN effect, reducing the minimum energy necessary to promote a carrier from the VB to the CB.

The BGN theory predicts [37] that concentrations above 10^{17} cm^{-3} would be sufficient to produce a measurable BGN effect using B and P, which lie well below the solid solubility limit of these impurities on Si [38], set in the order of 10^{20}–10^{21} cm^{-3}. Therefore, reducing the effective bandgap of Si using shallow dopants is relatively easy. The downside of the BGN is that the highest value that can be achieved, according to Klaassen et al. [37] is around 100 meV at the solid solubility limit of the impurities, which is insufficient to cover the SWIR range. Would it be possible to form an IB using deep centres, instead of shallow dopants? If so, the IB could lie between the VB and the CB, possibly located closer to the middle of the bandgap than when using shallow dopants.

In theory, if the impurity density is high enough, the same wave function overlapping process could take place with deep centres [39]. The minimum concentration level required to produce an effective overlapping between the discrete deep levels is called the insulator-to-metal transition limit [36, 40], among which the Mott transition is usually referred [40–43]. However, there is a fundamental difference between shallow impurities and deep centres: the insulator-to-metal transition limit [40] requires concentration levels several orders of magnitude higher than the solid solubility limit for most deep centres [44]. Therefore, technologically speaking, it is considerably more difficult to produce a deep IB than a shallow IB.

Under the assumption of the IB formation, there could be not only one transition from the VB towards the CB. Effectively, as depicted in Fig. 1.3, carriers could promote from the VB to the CB through the IB via two different transitions, each of them with energy lower than the bandgap, marked as E_H and E_L in the figure. Given the new transitions with lower energy than the bandgap, the deep IB concept could be

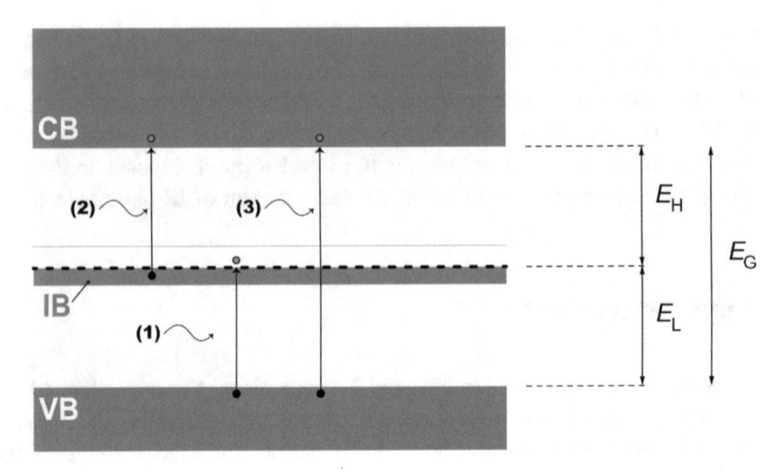

Fig. 1.3 Band diagram of a solar cell with an impurity band. The three possible transitions are labelled as (**1**), from the VB to the IB, (**2**) from the IB to the CB and (**3**) as the band-to-band transition from the VB to the CB

used to increase the absorption coefficient of the host semiconductor towards longer wavelengths.

In this thesis, we will focus on extending the Si photoresponse towards the SWIR range, aiming to answer the demands from both industry and research in this range of the IR. It is known that transition metals [44, 45] and some elements in the group of the chalcogens [46] (S, Se or Te) produce deep levels in silicon. Other elements in the III-V groups has been found to produce deep centres in Si, mainly P, B, In, Ga and As, although their practical applications are relegated to the very low temperature range [14]. In this last case, shallow dopants at cryogenic temperatures may form an IB very close to either the CB or the VB. If the temperature is low enough, impurities may not be ionised, making it possible to observe the optical transition between them. Besides, surface or bulk defects are known to produce discrete levels into the Si lattice, that could also enhance the sub-bandgap responsivity of Si-based devices [47]. In our research, we will focus on Ti, following the experience acquired in our research group since the thesis of J. Olea [48] and several published works [43, 49–58]. Ti exhibits three different deep levels when introduced into a Si lattice, according to Mathiot et al. [59]: two donor levels $E1 = E_C - 0.08$ eV and $E2 = E_C - 0.28$ eV and an acceptor level $H1 = E_V + 0.25$ eV, at concentrations below the solid solubility limit, set to around $10^{14} - 10^{15}$ cm^{-2}, according to the literature [60]. On the other side, other authors [42] theoretically estimated that the Mott limit (a particular case of the insulator-to-metal transition limit) is 6×10^{9} cm^{-2} for a generic impurity in a generic semiconductor. Furthermore, Pastor et al. [57]. experimentally determined that the IB formation limit of Ti atoms in Si could be in the order of 10^{20} cm^{-3}, which is close to the theoretical value. In this thesis, we will use the value of 6×10^{19} cm^{-2} as the limit of the impurity band formation, as it is the most commonly accepted value in the literature, knowing that the limit for Ti in Si should be close to that value.

In this text, reaching Ti concentrations higher than the IB formation limit will be referred to as "supersaturating". Hyperdoping is also used in several scientific manuscripts to refer to materials with impurity concentrations higher than the IB formation limit [46, 61–64]. Here, we choose supersaturating over hyperdoping as "doping" is preferably used to indicate electrical activation and substitutional occupancies, while supersaturating indicates that the impurity is in concentrations higher than the solid solubility limit and does not imply that Ti atoms are occupying substitutional positions (they seem to be mostly interstitial in the particular case of Ti) [48]. Please note that "supersaturating" is not strictly accurate in our case either, as the condition of supersaturation refers to the solid solubility limit, not to the IB formation limit, which might be higher.

Historically in our research group, the line of research dealing with Ti implanted Si had its origin on the works of A. Luque and A. Martí in 1997, where they applied the concept of the impurity band formation in the field of solar cells [42]. They stablished the basis of the Intermediate Band Solar Cell concept, enumerating a series of conditions that could lead to an important increase in the cell efficiency, overcoming the Shockley-Queisser limit of efficiency of single junction solar cells to reach up to 63%. However, the fabrication of a photodetector does not necessarily need to follow the same restrictions as those described by A. Luque and A. Martí. For example: in photodetectors, making use of both transitions is not necessary; using one of the two transitions will suffice.

However, the application of the deep IB concept into a real device faces several difficulties: the limit to supersaturate Si with Ti atoms is five orders of magnitude higher than the solid solubility limit, which rules out conventional thermal diffusion processes in order to avoid segregation or dissociation. Thus, the technique of choice to introduce Ti in soaring concentrations is the ion implantation, which is detailed in Sect. 2.1.2 Ion implantation processes allow the introduction of almost any isotope at any desired concentration, with the possibility to tune the profile by choosing the appropriate energy of implantation. The main downside is that the implantation process induces damage in the crystal lattice, in a first approach, proportional to the concentration of atoms implanted. Given the high concentrations required to supersaturate Si, the damage severely affects the crystalline structure, which may even result in partial amorphisation of the implanted layer. Damaged layers are not interesting from the point of view of photodetector fabrication; both the mobility and the carrier lifetime strongly depend on the crystal quality, and lower values could decrease the photoresponse of the device. Thus, monocrystalline Ti implanted layers are preferred for device fabrication.

In order to recover the crystal quality, thermal annealing processes involving high temperatures could provide enough energy to the atoms in the lattice to heal most of the damage. However, since we must avoid Ti segregation or agglomeration, we must seek for thermal annealing processes with short duration. In this field, laser processes in the range of femto- and nanoseconds gather all the desired characteristics to recover the crystalline quality, while being sufficiently short to minimise segregation or agglomeration of Ti atoms. More insight about laser annealing processes is given in Sect. 2.1.4. Other authors have recently used the Flash Lamp Annealing (FLA) technique, in the time scale of milliseconds, with satisfactory results [65, 66].

Hence, the supersaturation of Si using Ti, another relatively common element on the Earth's crust, offers promising characteristics to potentially answer all the questions that are currently unresolved in the world of SWIR range photodetectors. It is Si-based, the materials are abundant, non-contaminant and the techniques used to supersaturate Si are relatively common so they can be potentially integrated into a commercial CMOS route.

1.4 Photodetectors

Up to this point, we have chosen a goal, and we have briefly described how to pursue it, using Ti supersaturated Si layers. The next step is to decide the appropriate device that will advantage from the properties of this light-absorbent material [55]. Photodetectors are generally classified attending to which parameter is modified under illumination. The most common are described below:

Thermal detectors. They measure a change on the temperature of the so-called receptor element under illumination. The absorbed photons generate phonons inside the receptor, which in turn increase the temperature of the material. These were the first type of detectors used to detect the IR. In fact, the principle of operation is the same that W. Herschel used when he discovered the IR spectrum. The increase of the temperature caused by the illumination may change other physical properties of the detector. Pyroelectric detectors rely on crystalline receptors that exhibit anisotropic thermal expansion coefficients. Thus, after a change in the temperature, the crystalline structure may be tensioned. Like in piezoelectric materials, the tension can be translated to an electric field, which can be measured as voltage between two sides of the crystal. Other type of thermal detectors uses the thermocouple effect to measure the change in the temperature of any of the metals used in the junction as a voltage. Finally, the simplest form of thermal detectors would be those who monitor the resistance of the receptor, which may be temperature dependent. Devices using this principle are called bolometers. Thermal detectors are relatively slow, as they depend on how fast the receptor changes its temperature, but they feature broader bandwidth than other possibilities.

Photonic detectors. This type of detector relies on semiconductor materials to operate. The receptor, when illuminated, may absorb photons when their energy is higher than the bandgap of the semiconductor, promoting free carriers from the VB to the CB. The photogenerated carriers modify the conduction properties of the receptor. According to which conduction property is modified, we find two sub-divisions. Photoconductors are those where the carriers contribute to a change in the conductance of the receptor under illumination. Photodiode detectors, on the other side, operate similarly to a solar cell, by separating carriers using the electric field built inside a p-n junction. These detectors are usually more sensitive to light than their photoconductive counterparts. Photodiode detectors may work in three different regimes, depending on the polarisation of the device: it is possible to measure the short-circuit photocurrent, the open-circuit voltage or the photocurrent in reverse polarisation. In the first two cases, the device is said to be measured in photovoltaic mode. The application and the range of detection may determine which type of detector must be used.

There are several parameters that quantify the performance of a detector. Due to the big differences in the modes of operation of each type of detector, there is a consensus where the specific detectivity D^* is used to compare between different technologies [67, 68]. D^* is a figure of merit that takes into account the area of the sensor, the noise level at a certain bandwidth and the responsivity of the detector. Higher specific detectivity values are preferred at the range of interest. The next graph shows the specific detectivities and the range of operation of the most common technologies, available nowadays, that operates at room temperature in the NIR and SWIR range.

If we could fabricate Ti supersaturated Si photodetectors with QE in the order of 0.1% at 1.5 μm, while keeping the same noise level as in the sample labelled as Si in Fig. 1.4, the specific detectivity at said wavelength could be in the order of 10^{10} Jones, relatively close to the values of InGaAs photodetectors. Even if the Ti supersaturated Si photodetector had lower specific detectivity values, somewhere around 10^8 Jones, it could still be potentially interesting from the commercial point of view: it would displace part of the applications of microbolometers, provided that photonic detectors are usually faster, exhibiting higher cut-off frequencies. Besides, if the Ti supersaturated material could be integrated in a CMOS route, it would lead to a considerable decrease on the price per pixel as compared to microbolometers (see Fig. 1.2). Thus, although QE as high as possible are preferred, reaching QE levels under 0.1% could represent a potential advantage over other already commercially viable alternatives.

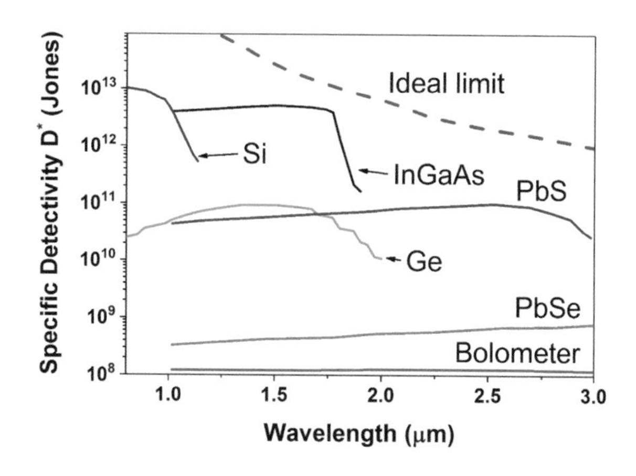

Fig. 1.4 Specific detectivity of several technologies, operating at room temperature, commercially available in the NIR and SWIR range. Data taken from Rogaslki [23]

1.5 Pixels and Image Formation

Most of the applications require the formation of images in the IR range. The first attempts to form images using semiconductors as sensing materials date from the 1960s, and, as many other technological advances, were a result of the necessity of the military industry, when they wanted to take aerial pictures from satellites [69]. The first pictures taken from satellites were performed using photochemical film rolls, that were launched from the satellite to the atmosphere and later retrieved for developing, under the CORONA missions [69]. The logistics and cost of operation were prohibitive, so they designed an integrated development system inside the satellite. The developed film was later scanned by an array of photodetectors, which translated the picture into an electrical signal that could be sent to the ground control centre [70]. Later, the first Focal Plane Arrays (FPA) were developed, in which several small sensing elements (**Pict**ure **El**ements, or pixels [71]) were arranged in a matrix. A focusing systems would form an image onto the plane containing the pixels. The electrical signal coming from each pixel is read, and a processing hardware and/or software translates each individual signal to a grey scale, which is later displayed in a monitor.

Depending of the structure and the functioning principles, FPAs based on photonic devices can be divided into two main groups.

<u>Charge-Coupled Devices</u> (CCD). The first developments of FPAs were based on CCD technology [72]. It was invented in the AT&T Bell laboratories by W. Boyle and G. E. Smith in 1969 [72]. A Metal-Oxide-Semiconductor (MOS) transistor generates carriers when it is illuminated. The amount of photogenerated charges is proportional to the intensity of light and the time it is illuminated. The chargers are stored inside the transistor and later transferred to the proper CCD part of the pixel. The stored charges from pixels within the same row/column are read one by one by the CCD (which acts as a horizontal shift register), which later amplifies the signal of the whole row/column and sends it to the Analog-to Digital Converters (ADC). The digital value is later analysed by the Read-Out Integrated Circuitry (ROIC). The first commercial digital cameras that were released to the public using this technology date from the beginning of the 1990s [73]. CCD sensors are characterised by their low noise, rendering good quality images, but most of the fabrication steps are specific of the technology (for example, the CCD part, the horizontal shift register system which transfers the chargers towards the ADC, is not compatible with CMOS routes), which is translated to more expensive sensors than competing technologies. Besides, the charge transfer process usually makes them slower. They also feature higher energy consumption.

<u>Complementary Metal Oxide Semiconductor</u> (CMOS). The absorption of light is similar as in the case of CCDs. A MOS transistor or a PhotoDiode (PD) generates charges as a consequence of the absorption of light, which are stored inside the bulk of the pixel. The charges are transferred to another MOS transistor, which converts the electrical charges into a voltage, which is later amplified, inside the same pixel [74, 75]. The signal is read by row or column selector transistors, which is sent to

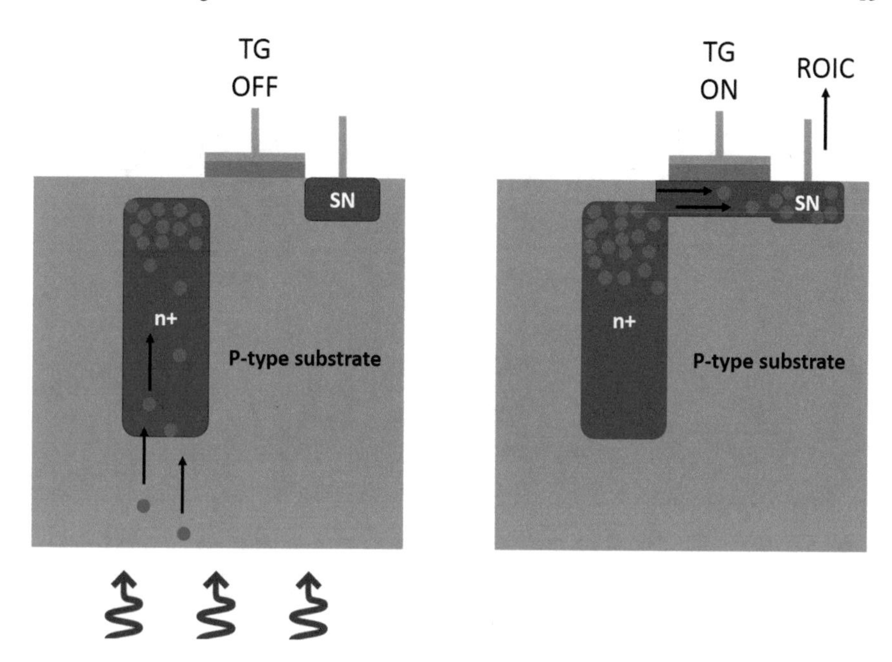

Fig. 1.5 Cross-section schematic of a Back Side Illuminated pixel while illuminated (left) and when reading the collected charges (right). When illuminated, the Vertically Pinned-Photodiode (V-PPD) stores charges in the upper part of the N^+ well, which is achieved by gradually doping the PD. When a positive bias is applied to the Transfer Gate (TG) an inversion channel communicates the V-PPD with the Sensing Node (SN), which communicates with the rest of the ROIC

the ROIC. A schematic of the basic principle behind a Back-Side Illuminated (BSI) CMOS pixel is shown in Fig. 1.5. The potential distribution and the motion of carriers is represented in Fig. 1.6.

The main difference between CCD and CMOS FPA detectors is the place where the charge is converted to voltage and amplified. In CMOS, each pixel has embedded the transistors that converts charges to voltage and amplifies it, which considerably speeds up the process of information transmission towards the ROIC. Depending on the pixel structure, there could be from 1 to 5 transistors per pixel on CMOS sensors, while in CCD sensors the signal is converted after the horizontal shift register, and it is converted for each row/column at the same time. However, as all transistors are based on CMOS technology, the fabrication process is considerably simpler than in the case of CCDs. The main advantages of CMOS Image Sensors in the early years of development were the lower price and lower power consumption, although they provided images of much lower quality than CCDs. As the technology has evolved in the last ten to fifteen years, CMOS Image Sensors have become the predominant technology for digital imaging in almost all applications. The best example is how any of us can have a 40 megapixels' camera inside a cell phone that is easily less than 8 mm thick. In recent years, and taking advantage of the Artificial Intelligence and machine-learning algorithms, the quality of the image produced by low-cost CMOS

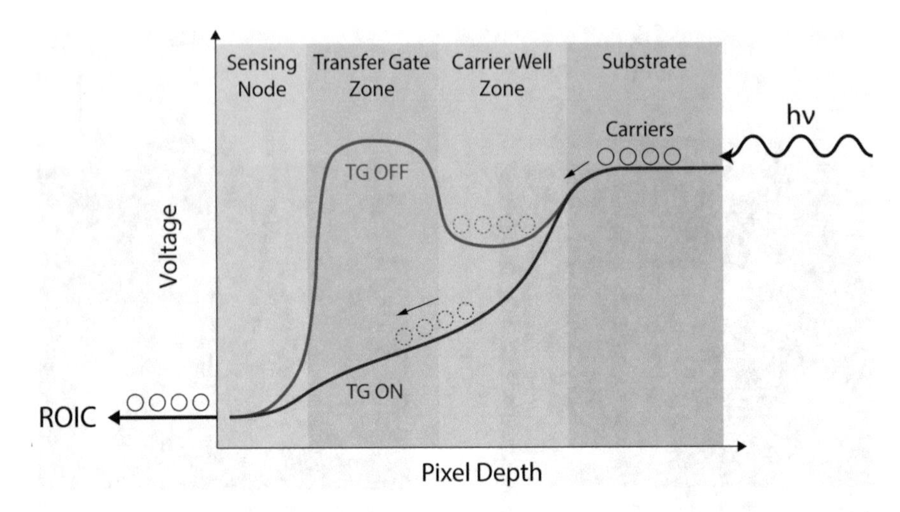

Fig. 1.6 Potential distribution inside a Back-Side Illuminated pixel when the Transfer Gate (TG) is on and off. Carriers are trapped in the N^+ well while the TG is off. After TG is activated, the carriers flow towards the SN, which converts the stored charges to voltage through another MOS transistor, which sends the signal towards the ROIC

sensors is outstanding, as compared to what we had less than 10 years ago. Therefore, CCD sensors nowadays are relegated to the use in astronomy sensors or in high-end cameras, where price and size are not an issue.

Therefore, the development of the Ti supersaturated Si material will be oriented towards the integration within a CMOS Image Sensor, as it is the most widely used imaging technology nowadays.

1.6 STMicroelectronics

STMicroelectronics is the largest semiconductor chip maker in Europe, and, according to the latest analysis performed by Gartner [76], the 9th largest chip manufacturer in the world in 2018. Its origin is a consequence of the merge of two public companies in 1985: SGS Microelettronica, from Italy, and Thomson Semiconducteurs, from France.

The core of the company in Europe is located in Crolles (France), although the headquarters are located in Geneve (Switzerland). In Crolles, they have two separated modules: the 200 mm part, where they work exclusively on 200 mm wafers, production oriented, and the 300 mm part, where they use bigger wafers (300 mm), mainly focused on Research and Development. The company has focused its R&D efforts in the last years to provide solutions in smart mobility and in the IoT, from RF communication (low consumption RFID emitters and receivers, Wi-Fi connection) to imaging in the visible and IR, among many other applications as power electronics, home automation or small computerised solutions.

The collaboration with STMicroelectronics has been a fundamental part of this thesis. We contacted the company after an invited speech offered by one of their experts in Si photonics on a Summer Course in Gant (Belgium). After the first contact, we briefly defined the collaboration project, which was carried out in the frame of a six-month internship, financed by the Spanish Ministry of Science and Universities, under grant number EEBB-I-17-12315. The collaboration project, where the Ti supersaturated material was implemented into a CMOS pre-commercial route, is described in Chapter 5.

1.7 Aim and Structure of the Thesis

The aim of the thesis is to fabricate and characterise photodiodes based on Si substrates with improved responsivity in the sub-bandgap region, that is, in the SWIR part of the spectra, up to 3 μm (0.45 eV) at room temperature. There is an increasing demand on cheap and fast IR photodetectors for the near-future challenges of the current society and an IR Si-based photodetector could potentially satisfy the expected demand, given that Si CMOS technology is relatively cheap, is abundant and with limited impact on the environment at the end of its lifecycle. Since bare Si is transparent in the SWIR range, we will apply the concept of materials with an Impurity Band through the concept of supersaturation, to increase the absorption coefficient of crystalline Si substrates towards the SWIR range. To achieve it, we will use a combination of an ion implantation process of Ti atoms at very high doses, combined with the use of a Nanosecond Laser Annealing (NLA) process to recover the crystal quality lost after the first step. This thesis continues the work described in the thesis of our co-worker E. García-Hemme [77], where he demonstrated that sub-bandgap photoresponse at room temperature was possible using supersaturated Si.

In Chapter 2, we describe the different experimental techniques used in this thesis to fabricate and characterise the Ti supersaturated Si layers, the devices manufactured using the supersaturated material and also part of the characterisation performed during the internship in STMicroelectronics.

In Chapter 3, we briefly describe some research done at material level. Later, we describe the design process of the pixel structure and the final layout of the pixel matrix in our facilities in UCM. In the second half, we describe the fabrication process of pixel matrices and the ulterior characterisation, by means of optical and electrical measurements.

Chapter 4 deals with the fabrication and characterisation of a new set of samples, in which we used a XeCl laser with long pulse duration to recover the damage coming from the Ti implantation. After the characterisation of the supersaturated material under different Ti doses and laser conditions, we fabricated transversal photodiodes, which were used to determine the best conditions for IR sensing using the XeCl laser.

Chapter 5 contains the research performed during my internship in STMicroelectronics. First, we describe the aim of the project and the steps used to integrate the Ti supersaturated layer into a pre-commercial CMOS route. Later, we followed the material fabrication route and the split plan of the available wafers. Then, we show the material characterisation. Finally, results of the fabricated CMOS Image Sensors using die-to-die and Wafer Level Characterisation are discussed.

In Chapters 6 and 7 we sum up the conclusions and milestones of this thesis and future works. Chapter 8 collects additional information that is not necessary to understand the core of the thesis but may be helpful to expand the knowledge of the reader. Lastly, while chapter 9 collects the contributions as peer-reviewed manuscripts and in national and international conferences coming from this thesis.

References

1. Grundmann M (2021) *The physics of semiconductors.* Springer International Publishing. ISBN: 978-3-030-51568-3
2. Neamen D (2003) *Semiconductor physics and devices.* McGraw-Hill. ISBN: 978-0-072-32107-4
3. Sze SM, Kwok KNg (2007) *Physics of semiconductors devices*, 3rd edn. Wiley. ISBN: 978-0-471-14323-9
4. Hook JR, Hall HE (1991) *Solid state physics*, 2nd edn. Wiley. ISBN: 978-0-471-92805-8
5. Ashcroft NW, Mermin ND (1976) *Solid state physics.* Brooks Cole. ISBN: 978-0-030-83993-1
6. Busch G (1989) *Early history of the physics and chemistry of semiconductors-from doubts to fact in a hundred years.* Eur J Phys 10:254–264. Doi:10.1088/0143-0807/10/4/002
7. Emsley J (2003) Book review: *nature's building blocks: an A–Z guide to the elements.* Oxford University Press, Oxford
8. Neamen DA (1997) *Semiconductor physics and devices*, vol 3. McGraw-Hill, New York
9. Wolf S (2003) *Microchip manufacturing.* Lattice Press, pp 584
10. Friedrich J (2016) *Methods for bulk growth of inorganic crystals: crystal growth.* Reference Module in Materials Science and Materials Engineering, Elsevier
11. Reuters (2019) SiC and GaN power devices market size, technology, segmentation, global industry will Register 32.8%CAGR reach US$1780 million in 2019–2024. *Reuters* https://www.reuters.com/brandfeatures/venture-capital/article?id=100798
12. Holst A (2019). *Semiconductor sales revenue worldwide from 1987 to 2019 (in billion U.S. dollars).* Statista https://www.statista.com/statistics/266973/global-semiconductor-sales-since-1988/
13. Radziemska E (2003) *Thermal performance of Si and GaAs based solar cells and modules: a review.* Prog Energy Combust Sci 29:407–424. Doi:10.1016/s0360-1285(03)00032-7
14. Rogalski A (2012) *History of infrared detectors.* Opto-Electron Rev 20: 279–308
15. Hulstrom R, Bird R, Riordan C (1985) *Spectral solar irradiance data sets for selected terrestrial conditions.* Sol Cells 15:365–391
16. Miller JL (1994) *Principles of infrared technolog: a practical guide to the state of the art.* Springer 1: 523
17. Miller JL, Friedman EJ (2003) *Photonics rules of thumb*, 2nd edn. Spie Press Book
18. Imaging ES (2018). *Short-wave Infrared Imagery (SWIR). European Space Imaging* https://www.euspaceimaging.com/wp-content/uploads/2018/06/EUSI-SWIR.pdf, 4
19. Vatsia ML et al. (1972) *Night-sky radiant sterance from 450 to 2000 nanometres.* Army Electronics Command. Fort Monmouth, NJ. AD-750 609, 42

20. Voshell A, Dhar N, Rana MM (2017) *Materials for microbolometers: vanadium oxide or silicon derivatives.* Proceedings volume 10209, image sensing technologies: materials, devices, systems, and applications IV; 102090 M SPIE
21. Kadlec EA (2011) *Thermal detecting in the long wafe infrared and very long wave infrared regions.* PhD dissertation, 67
22. Yarris L (2003) *Berkeley lab far-infrared detectors in orbit. science Beat berkeley labs.* https://www2.lbl.gov/Science-Articles/Archive/SB-MSD-SIRTF.html
23. Rogalski A (2002) *Infrared detectors: an overview.* Infrared Phys Techn 43:187–210
24. Market Study Report LLC (2019) *SWIR cameras market size analysis by growth application, segmentation and forecast to 2025.* https://www.marketwatch.com/press-release/swir-cameras-market-size-analysis-by-growth-application-segmentation-and-forecast-to-2025-2019-03-13
25. Yole Développement Reports (2016) *Uncooled IR imaging industry: the market is taking off.* http://www.yole.fr/UncooledIR_MarketOverview.aspx#.XW7eAy4zaUk
26. https://www.digitalglobe.com/products/short-wave-infrared
27. Ngo HT, Tao L, Zhang M, Livingston A, Asari VK (2005) *A visibility improvement system for low vision drivers by nonlinear enhancement of fused visible and infrared video.* IEEE Computer Society Conference on Computer Vision and Pattern Recognition
28. Haas H, Yin L, Wang Y, Chen C (2016) *What is LiFi?* J Light Technol 34:1533–1544
29. Visible Light Communication. *What is visible light communication?* http://visiblelightcomm.com/what-is-visible-light-communication-vlc/
30. Alkholidi AG et al. (2014) *Free space optical communication—theory and practices.* IntechOpen 55
31. TechInsights (2017) *Cost comparison—Huawei mate 10, iPhone 8, samsung galaxy S8.* https://www.techinsights.com/blog/cost-comparison-huawei-mate-10-iphone-8-samsung-galaxy-s8
32. Dhariwal SR, Ojha VN (1982) *Band-gap narrowing in heavily doped silicon.* Solid State Electron 25:909–911
33. Lowney JR (1985) *Band-gap narrowing in the space-charge region of heavily doped silicon diodes.* Solid State Electron 28:187–191
34. Mnatsakanov T., Pomortseva LI, Yakovlev DG (1994) *Estimate of the effective narrowing of the band-gap in heavily-doped layers of silicon structures.* Semicond 28:1059–1061
35. Matsubara T, Toyozawa Y (1961) *Theory of impurity band conduction in semiconductors.* Prog Theor Phys 26:739–756
36. Mott NF, Twose WD (1961) *The theory of impurity conduction.* Adv Phys 10:107–163
37. Klaassen DBM, Slotboom JW, de Graaff HC (1992) *Unified apparent bandgap narrowing in n- and p-type silicon.* Solid-State Electronics 35(2)
38. Jones SW (2008) *Diffusion in silicon.* IC Knowledge, LLC
39. Shalimova KV (1985) *Physics of semiconductors.* Energoatomizdat, Moscow
40. Mott NF (1968) *Metal-insulator transition.* Rev Mod Phys 40(7)
41. Belitz D, Kirkpatrick TR (1994) *The Anderson-Mott transition.* Rev Mod Phys 66:261–380
42. Luque A, Marti A (1997) *Increasing the efficiency of ideal solar cells by photon induced transitions at intermediate levels.* Phys Rev Lett 78:5014–5017
43. Olea J et al. (2010) *High quality Ti-implanted Si layers above the Mott limit.* J Appl Phys 107
44. Schibli E, Milnes AG (1967) *Deep impurities in silicon.* Mater Sci Eng 2:173–180
45. Mott NF (1949) *The basis of the electron theory of metals, with special reference to the transition metals.* Proc Phys Soc *A* 62(7)
46. Newman BK, Sher M-J, Mazur E, Buonassisi T (2011) *Reactivation of sub-bandgap absorption in chalcogen-hyperdoped silicon.* App Phys Let 98:251905
47. Casalino M, Coppola G, Iodice et al. (2010) *Near-Infrared sub-Bandgap all-silicon photodetectors: state of the art and perspectives.* Sensors 10:10571–10600
48. Olea J (2009) *Procesos de implantación iónica para semiconductores de banda intermedia.* Thesis dissertation
49. Gonzalez-Diaz G et al. (2009) *Intermediate band mobility in heavily titanium-doped silicon layers.* Sol Energ Mat Sol C 93:1668–1673

50. Olea J, Gonzalez-Diaz G, Pastor D, Martil I (2009) *Electronic transport properties of ti-impurity band in Si*. J Phys D Appl Phys 42
51. Olea J et al. (2009) *High quality Ti-implanted Si layers above solid solubility limit*. Proceedings of the 2009 Spanish conference on electron devices, pp 38–41
52. Olea J, Pastor D, Martil I, Gonzalez-Diaz G (2010) *Thermal stability of intermediate band behavior in Ti implanted Si*. Sol Energ Mat Sol C 94:1907–1911
53. Pastor D et al. (2011) *UV and visible Raman scattering of ultraheavily Ti implanted Si layers for intermediate band formation*. Semicond Sci Tech 26
54. Olea J, Pastor D, Toledano-Luque M, Martil I, Gonzalez-Diaz G (2011) *Depth profile study of Ti implanted Si at very high doses*. J Appl Phys 110
55. Olea J, del Prado A, Pastor D, Martil I, Gonzalez-Diaz G (2011) *Sub-bandgap absorption in Ti implanted Si over the Mott limit*. J Appl Phys 109
56. Olea J et al. (2011) *Two-layer hall effect model for intermediate band Ti-implanted silicon*. J Appl Phys 109
57. Pastor D et al. (2012) *Insulator to metallic transition due to intermediate band formation in Ti-implanted silicon*. Sol Energ Mat Sol C 104:159–164
58. Olea J et al. (2012) *Low temperature intermediate band metallic behavior in Ti implanted Si*. Thin Solid Films 520:6614–6618
59. Mathiot D, Hocine S (1989) *Titanium-related deep levels in silicon—a reexamination*. J Appl Phys 66:5862–5867
60. Hocine SAMD (1988) *Titanium diffusion in silicon*. Appl Phys Lett 53:3
61. Ertekin E et al. (2012) *Insulator-to-metal transition in selenium-hyperdoped silicon: observation and origin*. Phys Rev Lett 108
62. Mailoa JP et al. (2014) *Room-temperature sub-band gap optoelectronic response of hyperdoped silicon*. Nat Commun 5
63. Franta B et al. (2015) *Simultaneous high crystallinity and sub-bandgap optical absorptance in hyperdoped black silicon using nanosecond laser annealing*. J Appl Phys 118
64. Yang W et al. (2017) *Au-rich filamentary behavior and associated subband gap optical absorption in hyperdoped Si*. Phy Rev Mater 1
65. Liu F et al. (2017) *Realizing the insulator-to-metal transition in Se-hyperdoped Si via non-equilibrium material processing*. J Phys D Appl Phys 50
66. Liu F et al. (2018) *On the insulator-to-metal transition in titanium-implanted silicon*. Sci Rep-Uk 8
67. Jones RC (1957) *Quantum efficiency of photoconductors*. Proc. IRIS 2
68. Jones RC (1960) *Proposal of the detectivity D* for detectors limited by radiation noise*. J Opt Soc Am 50:1058–1059
69. Baunmann PR (2009) *History of remote sensing, satellite imagery. Department of geography*, State University of New York College, Oneonta. http://employees.oneonta.edu/baumanpr/geosat2/RS%20History%20II/RS-History-Part-2.html
70. Vick CP (2007) *KH-11 reconnaissance imaging spacecraft*. Globalsecurity.org https://www.globalsecurity.org/space/systems/kh-11.htm
71. Graf RF (1999) *Modern dictionary of electronics 7th edition*, p 869. Newnes Elsevier
72. Janesick JR (2001) *Scientific charge-coupled devices*. Spie Press Book 1:920
73. Digital Kamera Musseum (2015) *Dycam Model 1 (1990)*. https://www.digitalkameramuseum.de/en/cameras/item/model-1
74. Fowler BLX, Vu P (2006) *CMOS image sensors—past present and future*. Society for imaging science and technology ICIS '06 international congress of imaging science, 8
75. Theuwissen AJP (2008) *CMOS image sensors: State-of-the-art*. Solid State Electron 52:1401–1406
76. Gartner (2019) *Gartner says worldwide semiconductor revenue grew 13.4 percent in 2018; increase driven by memory market*. https://www.gartner.com/en/newsroom/press-releases/2019-01-07-gartner-says-worldwide-semiconductor-revenue-grew-13
77. Garcia-Hemme E (2015) *Respuesta infrarroja en silicio mediante implantación iónica de metales de transición*. Thesis dissertation

Chapter 2
Experimental Techniques

This section deals with the experimental techniques that have been used during the thesis. It covers several areas, from material to device preparation and characterisation. At each subsection, we briefly describe an experimental technique, explaining also its utility in our research.

2.1 Material Preparation

Every sample or fabricated device requires a specific preparation method, which transforms the raw material, in our case crystalline silicon, into the desired device. This section describes how we modify the supplied Si substrate into the Ti supersaturated material used for sample fabrication.

2.1.1 Silicon Substrates

The starting material for all the samples and devices analysed in this thesis is monocrystalline silicon, which is commercially available in the form of planar wafers, featuring different thicknesses, diameters, crystal orientations and electrical properties.

We have used mainly 2” (50.8 mm) diameter wafers, which are suitable for all the equipment located at Universidad Complutense de Madrid (UCM) laboratories. However, all the samples used in the internship done at STMicroelectronics in France were based on 300 mm diameter wafers. Although we have used both p-type and n-type wafers, almost the entire work relies on p-type wafers. Ti implanted silicon on n-type substrates has already been widely studied in our research group, in particular

D. Montero Álvarez, *Near Infrared Detectors Based on Silicon Supersaturated with Transition Metals*, Springer Theses,
https://doi.org/10.1007/978-3-030-63826-9_2

Table 2.1 Wafer properties for each crystalline Si substrate used in this thesis. SSP stands for Single-Side Polished, and DSP for Double-Sided Polished

Substrate Code	Diameter (mm)	Thickness (μm)	Type	Resistivity ($\Omega \cdot$cm)	Crystal orientation	Surface polishing
P1	50.8	300 ± 25	p	1–10	<100>	SSP
P2	50.8	300 ± 25	p	1–10	<100>	DSP
P3	300.0	750 ± 50	p	3	<100>	DSP
N1	50.8	300 ± 25	n	1–10	<100>	DSP
N2	50.8	300 ± 25	n	200	<111>	DSP

by J. Olea [1] and E. García-Hemme [2] in their respective thesis and scientific contributions.

Details on each of the substrates used in this work are displayed in the next Table 2.1.

The wafers are usually identified in the backside by diamond pencil marking or laser scribing a reference code. After, they are cleaned in acetone and IsoPropyl Alcohol (IPA) to remove solid particles that could have been deposited after the marking process.

2.1.2 Ion Implantation Process

The first successful experiments using ion implantation processes date from 1954, when W. Shockley patented it for the first time, titled "Forming semiconductive devices by ionic bombardment" [3]. The purpose of the ion implantation process is to change the chemical, electrical, structural or optical properties of a target (in our particular case, silicon wafers). It is a low thermal budget process that is widely used in the industry of microelectronics, material fabrication and energy harvesting. The most common use for this technique is to dope semiconductors, aiming to mainly change the electric properties of the implanted layer. It is mostly used at room temperature, although there are some applications that require low temperatures [4].

The basic principle behind the ion implantation process is the isolation of atoms of a certain specie (being possible to even select the desired isotope) coming from a solid, liquid or gaseous source, that are later accelerated towards the target, at sufficiently high energy to introduce or implant them in the atomic structure of the target.

In the world of microelectronics, ion implantation doping processes are usually preferred over thermal diffusion processes due to their high surface selectivity and lower energy consumption as compared to the latter [5]. In theory, thermal diffusion may not be suitable to produce supersaturated materials, as the process is performed in thermodynamic equilibrium, so concentrations higher than the solid solubility limit of the impurity cannot be attained. In contrast, ion implantation allows the introduction of a soaring amount of dopants in the host material at room temperature, even several orders of magnitude higher than the solid solubility limit of the impurity [6]. This fact makes the ion implantation an ideal candidate to fabricate supersaturated materials. However, this will induce a substantial amount of defects in the lattice of the host material, caused by the inelastic collisions between the dopants and the host atoms, usually amorphising the implanted layer [7, 8]. This damage may be healed in the next fabrication processes by providing energy to the lattice, usually in the form of heat, like furnace or laser annealing steps. The NLA process may also be used to modify the relative position of the implanted atoms with respect to the lattice: from interstitial to substitutional positions [9, 10]. However, for Ti atoms, it has been demonstrated that most of the implanted atoms remain in interstitial positions, even after the NLA process [6].

There are three main ion implantation parameters:

Dose. It indicates the amount of implanted atoms per unit area, and it is expressed in atoms per square centimetre (cm^{-2}). It is the parameter most related to the change in the target properties (mainly conductivity in the case of semiconductor targets).

Energy. The ions need a certain amount of kinetic energy, in order to penetrate the atomic lattice of the target. Increasing the energy of the implanted atoms pushes the atoms deeper into the lattice, producing in the end thicker implanted layers.

Tilt angle. It is possible to tune the angle between the normal vector of the target and the incident trajectory of the ions. We usually set the tilt angle to 7°, which may significantly reduce the channelling effect on Si substrates [11].

Dose and energy parameters determine the dopant profile distribution. Numerical simulations are usually run to determine more precisely the projected range and their expected in-depth distribution. Although there are some analytic models that give an estimation of both the projected range and the dopant profile, namely Gaussian or Pearson-IV distributions [12], Technology Computer Aided Design (TCAD) programs are preferred. In this work, we have used three different programs, a freeware program called Stopping and Range Ions in Matter (SRIM), and two licensed programs: Athena, part of the Silvaco TCAD suite and Sentaurus, part of the Synopsys TCAD suite.

During the development of this thesis, we have used a Varian CF3000 ion implanter refurbished by Ion Beam Services (IBS) France, capable to use acceleration voltages from 20 up to 180 keV. It is located in the dependencies of the Centro de Asistencia a la Investigación (CAI) de Técnicas Físicas at the Physics Faculty of Universidad Complutense de Madrid (Fig. 2.1).

During this thesis, we have used several ion implantation processes. We have implanted up to five different elements with a Freeman ionisation source (Table 2.2):

In general, ion implantation on silicon substrates is performed to change the electrical properties of the implanted layer, mainly conductivity. However, not only electrical properties can be changed. It is possible to change the optical properties, as explained in the introduction chapter, by implanting transition metals or chalcogens. Several works have demonstrated that the absorption coefficient is strongly dependent on the implanted dose of these elements, among other parameters [9,

Fig. 2.1 Varian CF3000 ion implanter used in this thesis, located at CAI Técnicas Físicas UCM

Table 2.2 Source elements and isotopes used for ion implantation in this thesis

Element	Isotope	Precursor
Titanium	^{48}Ti	$TiCl_4$ (liquid)
Silicon	^{28}Si	SiF_4 (gas)
Boron	^{11}B	BF_3 (gas)
Phosphorus	^{31}P	P (solid)

13–16]. The ion implantation step is particularly important in our research, as it is the main experimental technique that allows us to produce supersaturated materials, particularly with Ti atoms.

In the case of 50.8 mm wafer production, ion implantation process is the second step performed after the previously described wafer marking process. Most of the wafers used were back implanted with either boron or phosphorus to assure a back ohmic contact (also known as Back-Surface Field, BSF). In the case of 300 mm wafer production, there are several tens of different ion implantation steps, used to fabricate the pixel structure and also the ROIC, comprising different elements, doses and energies. Ti ion implantation in those wafers is performed on the backside (illuminated side) after the last FEOL process.

In all cases, the resultant implanted layer is full of stacking faults, point defects and possibly amorphisation of the first surficial nanometres, depending on the ion implantation parameters [8]. The next fabrication step should be restoring the damage caused by the implantation. This is usually accomplished by using heat treatments of different nature, described in the next sub-sections.

2.1.3 Rapid Thermal Annealing Process

One of the most used thermal annealing processes in the microelectronic industry is the Rapid Thermal Annealing (RTA), which consists of a furnace capable of achieving high heating and quenching rates within a short interval of several seconds. The furnace features electrical resistive elements, which heat up when current passes through them. The furnace chamber is usually filled with gas, in our case, Ar, N_2 or a mixture of argon with 5–10% hydrogen, labelled "forming" gas, depending on the application. The latter has been a key gas in the development of the microelectronics industry as it is widely used to activate metal contacts [17], thus the name "forming". The maximum achievable temperature depends on the furnace model used, but they are usually in the order of 1100 to 1500°C. The heating rate could be in the order of 100–300°C/s, while quenching rates must be lower to avoid cracking or deformation on the wafers, in the order of 10–100°C/s. Quenching is usually achieved by using a cooled gas flow through the chamber. Typical applications involve RTA processes of around 30–120 seconds.

The equipment used in this work, located at CAI Técnicas Físicas in UCM, is a MPTC RTP 600 model, with the next characteristics:

- Maximum wafer size: 100 mm (4")
- Maximum temperature: 1400 K
- Heating rate: 0 to 150 K/s
- Quenching rate: 0 to 100 K/s

In our research, we have used RTA processes to activate mainly boron and phosphorus dopants. We have also used a longer thermal process, with forming gas, called Forming Gas Annealing (FGA), aiming to improve the contact resistance between the metal and the semiconductor [17]. It is also widely used to passivate the surface or defects of silicon samples [18]. It may comprise several minutes, while temperatures are in the order of 200–600°C.

2.1.4 Nanosecond Laser Annealing Process

Supersaturated materials, as it is the case of Ti supersaturated Si, the core material of this work, should be fabricated using out-of-the-equilibrium techniques. It is in this context when we consider the use of non-conventional annealing techniques, as the Nanosecond Laser Annealing (NLA).

As opposed to what we have previously described in Section 2.1.3, we are seeking for a brief thermal process, long enough to restore the damage caused by the ion implantation process but short enough to minimize the Ti dopant segregation. The reason behind this last affirmation is that Ti atoms have high thermal diffusivity in Si at high temperatures, especially in the liquid or molten phase [19], so they tend to out-diffuse if the melting time is sufficiently long. Nanosecond Laser Annealing (NLA) or Pulsed Laser Melting (PLM), equivalent terminologies found in the literature, seem to fulfil the previously described requirements, as the duration of the melting process is usually in the order of the pulse duration [1, 20–22].

The laser annealing process relies on the interaction of light and matter to produce heat in extremely short periods of time, going down even to the femtosecond scale. In the case of NLA processing of silicon substrates, lasers in the ultraviolet (UV) region are widely used in both research and industry. Although there are also NLA applications in the visible and in the IR part of the spectrum, UV laser are more common in the microelectronics industry. The reason behind the predominant use of UV lasers is due to the higher absorption coefficient of silicon at these wavelengths [23], which increases selectivity, as it would only heat up the first few microns. Most UV lasers feature wavelengths from 157 to 355 nm.

From the different commercially available laser configurations, excimer lasers are the most used NLA technologies in the silicon industry [24], with increasing importance as light sources for photolithography processes [25]. They rely on the relaxation process of previously excited groups of atoms called **exci**ted di**mers** (hence the **excimer** denomination) to produce monochromatic coherent light beams, which are focused at the output to increase the effective laser energy density (commonly referred to as "laser fluence"). The wavelength of the laser depends on the gases used in the resonant chamber: F_2 ($\lambda = 157$ nm), ArF ($\lambda = 193$ nm), KrF ($\lambda = 248$ nm), XeCl (308 nm) and XeF (351 nm). Each one of the possible combinations has its advantages and drawbacks [26].

Silicon absorption coefficient in the UV range is higher than 10^6 cm^{-1} according to M. Green [23]. Using the Beer-Lambert law, despising reflectance and assuming normal ray incidence, it would lead to the absorption of 99% of the incoming laser within the first 50 nm of implanted Si. This is extremely important and useful in microelectronics. It could be used, for example, to activate surficial implantations without affecting the previous doping processes, sensible to heat treatments, like in the activation of drain-source dopants of MOS transistors [27, 28]. This is a fundamental difference with respect to RTA, where the whole wafer had to be treated at the same temperature. Moreover, with the use of opaque masks, different areas of the device can be selectively exposed, leaving the covered areas unchanged.

Another aspect of heating up only a small fraction of the semiconductor wafer is the relatively rapid cooling process, which is the key to produce supersaturated materials. Wafers are several hundreds of microns thick. With NLA processes, only the first hundreds of nanometres are heated up, even reaching the liquid phase due to the high temperatures achieved, if the laser fluence is high enough. The rest of the wafer is still in solid phase, which could act as a nucleation centre. Since the thermal mass of the solid wafer is way higher than the small molten fraction, the cooling down process of the molten layer is expected to be very fast, about in the same order of magnitude as the NLA pulse duration. Silicon has higher heat diffusion coefficient than air, so heat is absorbed on the rest of the wafer rather than being transmitted to the atmosphere [29]. The resulting process is a molten layer that starts to solidify from the substrate, moving towards the surface at velocities, for silicon substrates, in the order of several metres per second [21]. Implanted atoms have little time to diffuse, so they are usually trapped inside the host semiconductor during the solidification process [30].

In this thesis, we aim to use laser fluences high enough to melt past the amorphous-crystalline interface, produced during the ion implantation process. This process is framed in the so-called liquid-phase epitaxy. During the melting process, there is a liquid layer atop of the solid substrate. It is known that diffusivity and solubility limit of an impurity is substantially higher in the liquid phase than in the solid phase [31]. Therefore, after the recrystallization process an impurity redistribution is expected with respect to the as-implanted distribution [32]. This is especially important with Ti, where extremely low solid solubility limits are measured experimentally. Ti atoms tend to stay in the liquid phase, thus following the solidification front towards the surface. This effect has been previously measured and described in other references [1, 13], labelling the Ti redistribution process as the "snow-plow" effect due to Ti atom migration towards the surface.

NLA process is not a straight-forward technique. There are plenty of parameters that need to be carefully monitored and studied in order to optimise the process and the properties of the resultant material. The melting and subsequent solidification processes are strongly dependent on the crystalline properties of the semiconductor lattice. Since we used an ion implantation process to supersature the Si lattice, there is an associated damage on the crystal lattice. Thus, parameters of the ion implantation, as the specie, the energy and the dose may play an important role in the thermodynamics of the solidifcation process [33, 34]. The complexity of the NLA process is especially increased in the case of supersaturated materials due to their particular properties, mainly very high doses, along with significant amounts of damage induced in the lattice. Fabricating a supersaturated material is a challenging task from the experimental point of view.

During our research, we have used two different excimer laser instruments, featuring different gases.

2.1.4.1 KrF Excimer Laser of IPG Photonics

We have been historically collaborating with IPG Photonics, a company located in New Hampshire (USA) since the beginning of our Impurity Band material research line. They offered the access to an IX-260 ChromAblate KrF excimer laser equipment. Its main characteristics are shown below:

- Laser pulse duration: 25 ns at Full Width Half Maximum (FWHM)
- Pulse shape: square-shaped spot of 1 mm^2 size.
- Pulse fluence: from 0.1 up to 2.0 J/cm^2.
- Uniformity: better than 2% by using a fly-eye filter
- Maximum allowed wafer size: 150 mm (6")

Since samples are usually bigger than the laser spot size, the stitching method is used, in which several laser shots are fired while moving the wafer in both x and y axis using a motorised and automatized stage. The overlapping between pulses was set to 10 μm.

This laser was used in the first stage of the thesis. Unfortunately, the company closed its research line on excimer lasers. Only a few of the 50.8 mm wafers used in this thesis were manufactured using this laser equipment.

2.1.4.2 XeCl Excimer Laser of Screen-Lasse

A thorough research was performed after the end of our collaboration with IPG Photonics. The start of the collaboration with STMicroelectronics France, within the framework of the internship of this thesis, led us to find a suitable laser with 300 mm wafer handling capabilities. The best candidate was a French supplier, Screen-Lasse, located in Paris. Aside from the wafer handling possibilities of this equipment, we were searching as well for a longer pulse duration laser. Some studies linked the longer pulse duration to longer melting durations, producing in the end epitaxially regrown layers with higher quality. Longer melt durations may allow atoms to have more time to be reorganized into a crystalline structure, thus having lower density of defects once the layer has solidified [22], as seen in previous works [28, 35]. However, the melting time should not be too long to avoid Ti diffusion and segregation.

The equipment used is an LT3100 Screen-Lasse with the next characteristics:

- Laser pulse duration: 150 ns at Full Width Half Maximum (FWHM).
- Pulse shape: square-shaped spot with two possible sizes, 10×10 mm^2 and 15×15 mm^2.
- Pulse fluence: from 1.0 up to 9.0 J/cm^2 for the small spot and from 0.8 to 3.6 J/cm^2 for the bigger spot.
- Uniformity: better than 2% by using a focusing set-up and a top-hat mask.
- Maximum allowed wafer size: 300 mm (12”).

The additional advantage of this laser set up is its bigger laser spot as compared to other available lasers, avoiding the use of “stitching” patterns to cover bigger areas. Some authors observed that laser stitching patterns increased the dark current of laser annealed devices [36], especially when the laser spot is smaller than the sensor size. When the laser spot size is bigger than the sensor it is called full die exposure. This laser equipment has annealed most of the wafers used in this thesis.

2.1.5 Preparing Samples: Cleaving Wafers

Wafers are useful because of their manageability, but their size make them unsuitable for a large number of experimental techniques or fabrication steps.

Research 50.8 mm wafers were manually marked and cut, using a Karl-Süss KG 502200 wafer cutter. After the cutting process, we marked the fragments on the backside with an identification code. Production 300 mm wafers were laser scribed by automated equipment.

2.2 Compositional and Structural Characterisation

In the previous sub-section, we described the material fabrication process. Here, we explain the different techniques used to characterise the material, from the compositional and structural point of view.

Between the raw material fabrication and the fabrication of prototype devices, it is necessary to characterise the material, in order to find out if it meets the expected properties:

- Ti concentrations higher than the IB formation limit to potentially form an impurity band.
- Good crystal quality in the implanted layer.
- Light absorption at wavelengths lower than the bandgap.
- Carrier generation coming from sub-bandgap absorbed light.

In this part of the thesis, we describe the experimental procedure to answer the first two items on the previous list. The third and fourth items are treated in Sect. 2.4.

2.2.1 Secondary Ion Mass Spectroscopy

After the fabrication process of any supersaturated material, the first thing to check is if the supersaturating condition has been achieved, by obtaining the dopant distribution profile as a function of the depth. Among the different available techniques, we are seeking for a technique offering the best in-depth resolution, with enough precision to go at least at the part per million (ppm) element detection. The most used technique is the Secondary Ion Mass Spectroscopy (SIMS).

SIMS technique uses an ion beam oriented towards the surface of the sample. The ion bombardment expels atoms from the surface (a sputtering process) that are later collected by a spectrometer, which identifies the incoming elements and their relative abundance as total ion counts versus the depth. We normalise the Ti signal by dividing it by the Si signal, to take into account the possible inhomogeneities in the extraction efficiency of the ions, especially in the first nanometres of the sample. In order to find out the impurity concentration value (depending on the equipment, a relative error of 5–10% is acceptable), the as-implanted profile is used as a reference [37] to calibrate the laser annealed curves. It is at this point when we can analyse if the material has been supersaturated by comparing the profile with the minimum required concentration value, the IB formation limit.

Two different SIMS equipment have been used for our samples. 50.8 mm wafers were analysed using a Time of Flight-SIMS (ToF-SIMS) equipment manufacture by IonTOF, ToF-SIMS[5] model located at Universidad de Badajoz (Spain). In all the measurements, the O_2 sputtering gun energy was set to 1 keV. Ti distribution profiles obtained on 300 mm wafers were measured using a Dynamic-SIMS (D-SIMS) equipment capable of measuring the profiles at wafer level. The equipment was a CAMECA IMS Wf located at STMicroelectronics SAS2 (Crolles, France). The sputtering gun was set to 3 keV.

2.2.2 *Energy-Dispersive X-Ray Spectroscopy*

Energy-Dispersive X-Ray Spectroscopy (XEDS, EDS, EDX or EDXS, according to literature) provides qualitative information about the chemical composition of the sample under study. The impinging electrons may excite inner electrons of the atoms in the surface of the sample. After the excitation and relaxation process, the atoms emit an electromagnetic radiation in the form of X-rays that are analysed by a spectrometer, which identifies the different elements present in the sample [38]. EDX equipment are usually integrated as modules into bigger electronic microscopes. The main advantage of this technique over SIMS is its high selectivity. Depending on the equipment, the electron beam can be narrowed down to several angstroms. SIMS only provides an average over a large surface of several square microns. On the opposite, EDX technique is only capable to detect atoms that are in an atomic concentration of at least 1–2%, while SIMS provided up to parts per million or even parts per billion, depending on the element [39].

EDX has been used in this thesis to identify the chemical properties of different layers studied mainly on cross-section TEM lamellas, like native oxide or Ti supersaturated layers, to mention some. The EDX module we have used in our research is an Oxford INCA, installed in a JEOL 3000F Transmission Electron Microscope.

2.2.3 *Transmission Electron Microscopy*

Transmission Electron Microscopy (TEM) is a characterization technique that uses high speed electrons to analyse a sample. Electron Microscopy techniques rely on the wave behaviour of highly accelerated electrons (in the range of keV), which are focused onto a sample. In the case of TEM, the electrons cross the sample, so transmitted electrons are analysed later by a variety of sensors. To do so, the sample must be sufficiently thin, in the range of several tens of nanometres, which requires specific sample preparation.

In our research, we used TEM technique to obtain information of the crystal quality and interface properties between the implanted layer and the substrate. Thus, we mainly analysed the transmitted electrons in TEM mode. We used a JEOL 3000F

TEM equipment, which is part of the specialised research facility "Infraestructura Integrada de Microscopía Electrónica de Materiales" (ELECMI) ICTS (Instalación Científico-Técnica Singular). It features a Schottky field emission electron gun with a maximum accelerating voltage of 300 keV. Its maximum resolution is 0.17 nm in TEM mode and 0.14 nm in STEM mode.

2.2.4 Raman Spectroscopy

Raman spectroscopy is based on the inelastic scattering of light and its interaction with matter. Monochromatic laser sources in the UV or visible range are mostly used. Atoms or molecules interact with the incoming laser light via phonons, vibrational or rotational states. After the interaction, the incoming photons interchange energy with the target, shifting its wavelength. Each Raman scattering process has a unique wavelength shift spectra that can be used to identify the elements present at the sample. As atoms in a crystalline lattice have different electronic structure as compared to amorphous materials, by examining the Raman spectra it is possible to identify also the crystalline quality of the sample [40]. We have used Raman spectroscopy in this work to further analyse the crystal quality of Ti supersaturated samples.

We have used a NT-MDT NTegra Spectra equipment featuring a Czerny-Turner monochromator with four different grating configurations to obtain the Raman shift, which belongs to the CAI de "Espectroscopía y Correlación" of the UCM. We used a green laser of 532 nm wavelength and 22 mW power output. The equipment features an optical system to focalise the laser beam into the sample down to less than one millimetre diameter spots. A motorised XY staged allowed to perform spatial scans along the surface of the sample. The minimum wavenumber resolution depends on the set-up, with a maximum of 0.1 cm^{-1}.

2.3 Surface Characterisation

Surface properties like structure, polycrystalline phases, roughness or simply device geometry are also important when fabricating devices. Thus, we used several techniques to characterise the surface of Ti supersaturated Si samples.

2.3.1 Scanning Electron Microscopy

Scanning Electron Microscopy uses high speed, focused electrons to sweep the surface of the sample and obtain information coming from the deviated electrons by the surface. Imaging is possible through this technique, as in previously described TEM technique. Secondary electrons, those generated from the sample by effect of the incoming radiation, are most commonly used for SEM detection and surface imaging, while back-scattered electrons, products of an elastic scattering process, are used for chemical element identification [41]. The main difference with respect to TEM is that samples do not need to be prepared in a specific way.

SEM micrographs have been used in this work to examine the surface topography, at both material and device characterization level. The equipment used, part of the ELECMI ICTS facilities, is a JEOL JSM 6335F, located at Centro Nacional de Microscopía Electrónica (CNME) in Madrid. It is capable to achieve a magnification of 500,000X with a maximum resolution of 1.5 nm.

2.3.2 Focused-Ion-Beam Scanning Electron Microscopy

Focused-Ion-Beam Scanning Electron Microscopy (FIB-SEM) is a sample preparation method used in combination with a SEM equipment. There are several applications using FIB-SEM, from lithography process or surface scribing up to TEM sample preparation. This technique uses a focused ion beam to erode the surface of the sample, also known as ion milling. However, FIB-SEM can be used as well to deposit materials atop the sample. By combining both capabilities of the FIB-SEM, TEM lamellas can be prepared using this technique [42], which is the main purpose of this technique within this thesis. Since we aim to find the crystalline quality of the implanted layer, we search for a cross-section lamella, which we obtained by means of FIB-SEM.

The FIB-SEM equipment used are a Helios Nanolab 600 and a Helios Nanolab 650, both belonging to the Laboratorio de Microscopía Avanzada (LMA) in Zaragoza, also part of the ELECMI ICTS. They feature a Ga^{+} ion gun to perform the ion milling process and a Pt source to deposit the protective layer.

2.3.3 Atomic Force Microscopy

Atomic Force Microscopy (AFM) is a technique that uses the interaction between the sample and a cantilever to measure different targeted parameters of the former. It is based on the Scanning Probe Microscopy (SPM) technique. The basic principle of this technique is to approach a highly sensitive cantilever to the surface of the sample. At sufficiently short distance, there is an electromagnetic interaction between the tip of the cantilever and the sample [43], which moves the tip. It is possible to map the surface as a function of different parameters, depending on the chosen experimental set-up. In this work we have used AFM in tapping mode as a technique to map the surface roughness of Ti supersaturated layers.

The equipment used is a Multimode Nanoscope III A manufactured by Bruker, located at the CNME, part of the ELECMI ICTS. Its resolution is better than 1 nm in tapping mode, the configuration used in this work.

2.3.4 Time-Resolved Reflectivity

Time-Resolved Reflectivity (TRR) technique provides information of the reflectivity of a surface as a function of time. It may be used in combination with annealing techniques, mostly laser annealing processes (like NLA, described in Sect. 2.1.4), where the melting process usually takes place in short periods of time, in the order of pico- or nanoseconds. The value of reflectivity is used as an indicator of the liquid or solid state of the surface, as it can be found in several works [22, 44]. In the case of silicon, its reflectivity is directly dependent on the temperature until the maximum value, which is reached in the liquid (molten) phase. Thus, with the TRR technique it is possible to obtain the average melting duration of a sample undergoing a laser process. Reflectivity is measured by pointing a laser source at the same surface and monitoring the reflected laser beam with a calibrated sensor as a function of time [45], at the same time the sample is being laser melted (with the use of a UV laser, as explained before). Another key characteristic of the TRR technique is that it is a non-contact, non-destructive measurement that can be performed at wafer level.

The TRR module we have used is part of the Screen LT3100 NLA equipment previously described in Sect. 2.1.4, located at Grenoble (France). It features a laser source with a wavelength of 632 nm with a time resolution of 2 ns.

2.3.5 Haze Measurement

Haze Measurement (HM) is a technique used to characterise the imperfections of a surface. Haze value is related to roughness, deposited particles, scratches or the presence of different crystal orientations in the surface [46]. It is determined by impinging a laser spot in glancing angle (usually 70°) onto the surface under study. A mirror, placed in the perpendicular of the point where the laser is focused, collects and redirects non-specular (diffuse) reflecting rays coming from the sample (see Fig. 2.2) towards two different sensors. The laser is swept across the whole wafer while the latter is rotating at high speed. The result is a Haze map of the whole wafer. By examining the Haze map, the possible nature of the roughness can be identified. If there are any particles or scratches present in the surface, they are easily identifiable [47]. Haze is expressed in parts per million (ppm) and the higher the value, the higher the deviation from a perfectly reflecting surface.

We used this technique in the thesis to identify the possible presence of polycrystals in the surface of laser annealed Ti supersaturated Si samples, which could help determine the best window process to achieve good crystal quality of the implanted layer. It is fast, non-contact and non-destructive, and it is also performed at wafer level.

We have used a KLA-Tencor Surfscan SP2 equipment available at CEA-LETI-Minatec campus in Grenoble (France). It uses a 355 nm laser source to determine Haze. It features a selecting mirror in order to choose the narrow or the wide sensor.

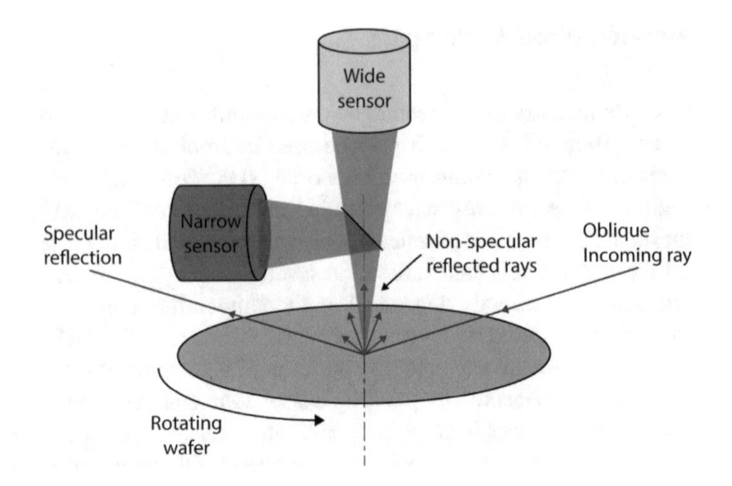

Fig. 2.2 Basic representation of the Haze Measurement procedure

The manufacturer sets a minimum detectable size particle of 30 nm using the oblique rays and wide sensor mode, which was the setup we used for the study present in this work. The Surfscan SP2 only accepts 200 and 300 mm diameter wafers.

Please note that Haze Measurement value will be referred to as "Haze" using the first capital letter to differentiate this technique from the regular noun "haze".

2.3.6 Optical Profilometry

Optical profilometry allows to measure the surface topography of the sample by using an image acquisition hardware, software-controlled focusing set up [48]. The software processes all the information and displays a three dimensional map of the sample. It is possible to obtain the size of the structures observed in the sample in the three dimensions as the whole system is calibrated.

We used this technique to study the morphology and the structure of Ti supersaturated layers, as well as a tool for quality inspection during device fabrication of pixel matrices, for example to check the lithography processes. The equipment used is a Wyko NT1100 Optical profilometer. The range (minimum discernible vertical distance) in VSI mode is around $\lambda/4$ of the light, which in our equipment is 632 nm, giving a maximum vertical resolution of around 160 nm.

2.4 Optical and Electrical Characterisation

Optical and electrical properties of the Ti supersaturated material are of extreme importance for the final device operation. Optical techniques were used to obtain the absorption properties of the Ti implanted layers, as well as in deposited dielectric layers, mainly silicon oxide and silicon nitride, during the device fabrication. From the electrical point of view, current-voltage characteristics may provide information about the electrical junction between the substrate and the Ti supersaturated material. Four point measurements, as in the van der Pauw configuration (resistivity or Hall effect) could provide information of the conduction processes of the Ti implanted layer. Finally, among all measurements, the photoresponse is the most important of all, as it is the experimental verification of the device performance under sub-bandgap illumination.

2.4.1 Transmittance-Reflectance Measurements

Transmittance-Reflectance (T-R) measurements are performed at wafer level on Ti supersaturated Si layers. It is possible to estimate the absorption coefficient, through the complex refraction index of the Ti implanted layer, by applying different theoretical models [49]. Transmittance and reflectance are two different measurements, usually performed on the same equipment but with different instrument configurations. The main purpose is to obtain both reflectance and transmittance, in our case at 8° for the former and at normal incidence for the latter, as a function of the wavelength. Simplifying the experimental set up, it consists of a monochromatic beam that impinges on a reference sensor and on the sample, which transmits or reflects the ray to another sensor. By comparing the signals from both sensors it is possible to extract the reflectance or transmittance value.

From Transmittance-Reflectance measurements, the absortance can be calculated as:

$$A(\lambda) = 1 - R(\lambda) - T(\lambda) \tag{2.1}$$

The absortance is the percentage of light that is not reflected nor transmitted in the sample. In the particular case of normal incidence this quantity is associated to the absorption processes inside the sample, value widely used in the literature [50]. The absortance value is one of the first calculated parameters after the fabrication of Ti supersaturated Si, as it is relatively easy and fast to obtain, in order to see if the material is absorbing photons with energy lower than the band gap. In ulterior analysis, we apply different models to both reflectance and transmittance to estimate the complex refractive index, which is later used to estimate the absorption coefficient, one key parameter of any semiconductor, but especially relevant on the study of supersaturated semiconductors.

During the thesis, we have used a Perkin Elmer UV/VIS/NIR Lambda 1050 Spectrophotometer, featuring an integrating sphere of 60 mm with an InGaAs sensor, capable to measure direct and diffuse reflectance. It offered the possibility to measure both reflectance and transmittance from 300 up to 2500 nm at room temperature. The equipment is part of the facilities of the Department of Energy of the Centro de Investigaciones Energéticas, MedioAmbientales y Tecnológicas (CIEMAT), in Madrid.

2.4.2 Optical Ellipsometry

Optical ellipsometry technique allows to estimate the thickness and complex refractive index of thin dielectric films, in our particular case, silicon oxide and silicon nitride layers. A laser source points a polarised beam towards the sample. The beam interacts with the sample according to its properties (mainly thickness, crystal orientation and complex refractive index), resulting in reflected and transmitted beams. The reflected beam is analysed, including its state of polarisation. By applying theoretical models, it is possible to estimate the properties of the dielectric layer [51].

Our experimental set up comprises a single-wavelength LSE-WS Stokes WAFERSKAN ellipsometer, which features a HeNe laser source (λ =632.8 nm) for both the alignment procedure and characterisation, the latter with an angle of incidence of 70°. A mounting motorised stage, with a precision of 10 μm, allows the mapping of the surface of any of the extracted properties of the layer (Fig. 2.3).

Fig. 2.3 LSE-WS Stokes Waferskan ellipsometer, located at the cleanroom of our research group (UCM)

2.4.3 Current-Voltage Characteristics

Current-voltage (IV) characteristics can provide information about the conduction mechanisms of the analysed sample. They are widely used in both research and industry environments as they are relatively fast and compatible with a wide selection of materials and structures. A bias is applied between (at least) two electrodes, which may produce a measurable current in the device [52].

Current-voltage characteristics have been widely used in this thesis, from material to device characterisation. Our experimental set-up is composed of several equipment:

- Keithley 2636A Source and Measure Unit (SMU). Mainly used to measure IV curves.
- Four-point station, manufactured by Everbeing. It consists of a black metallic box that contains four needles, to measure microscale devices. Features triaxial connections (Fig 2.4).
- Keithley SCS 4200 SMU. It contains four different programmable SMUs, which are used mainly for four-probe measurements, as in resistivity and Hall effect. It communicates and controls other instruments via GPIB ports, which allows the automation of electrical measurements.

2.4.4 Resistivity and Hall Effect: The van der Pauw Configuration

Resistivity is an intrinsic property of each material, which measures the degree of difficulty for an electrical current to pass through it. It is often measured in Ω·cm, and it can be calculated as follows [52]:

$$\rho = \frac{1}{q\left(\mu_n n + \mu_p p\right)} \tag{2.2}$$

Being ρ the resistivity, q the elemental charge of the electron, μ_n and μ_p the mobilities of electrons and holes, respectively, and n and p their volumetric concentrations.

L. J. van der Pauw demonstrated in 1958 that it is possible to obtain the resistivity and the Hall coefficient of a lamellae of arbitrary shape, provided the sample meet some criteria [53], using a four-probe configuration. Another advantage of this configuration is that it may minimise the effect of parasitic resistance elements, as the contact resistance or the influence of the cables [54]. In this thesis, we use square-shaped samples with one triangular contact at each corner.

In the van der Pauw configuration (Fig. 2.5 left), current is injected between two neighbouring contacts, while bias is measured in the remaining two. The sheet resistance can be obtained using the formulas:

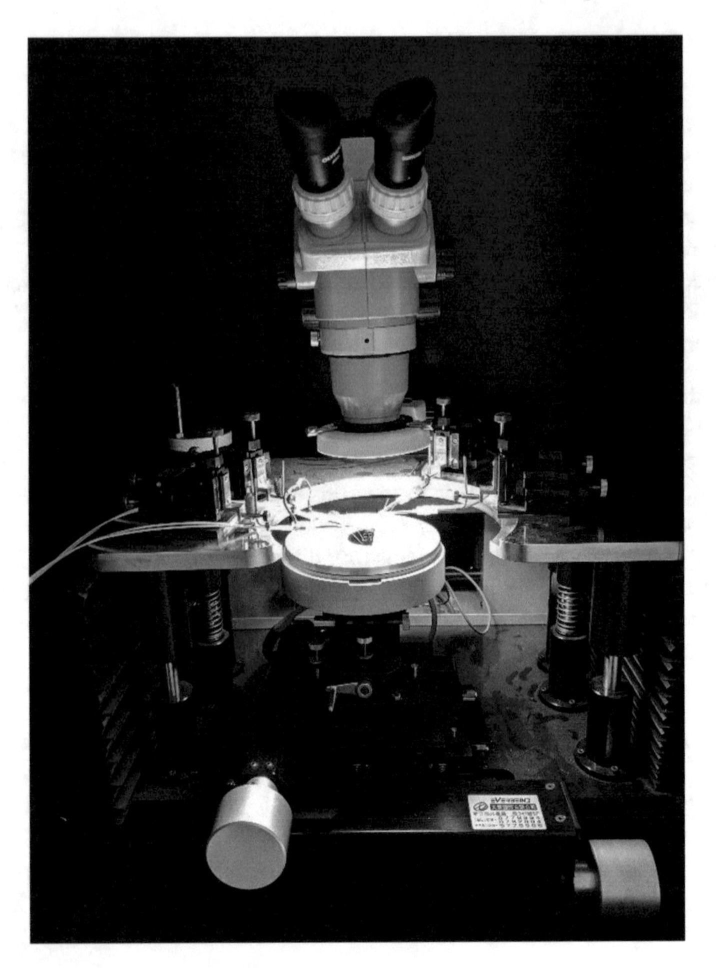

Fig. 2.4 Everbeing four-probe station

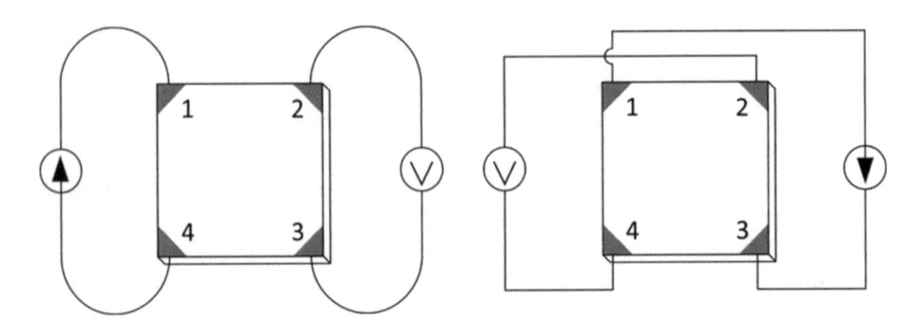

Fig. 2.5 Four-point measurements. Left: resistivity measurement. Right: Hall effect

$$R_s = \frac{\rho}{t} = \frac{\pi}{\ln 2}\frac{R_A + R_B}{2} f(Q) \quad (2.3)$$

$$Q = \frac{R_A}{R_B}; \; Q \leq 1 \quad (2.4)$$

where R_s is the sheet resistance, t is the thickness of the lamellae, Q is a geometric factor defined in Eq. 2.4, f(Q) is the van der Pauw function [53], strongly dependent of the geometry of the sample and R_A and R_B are the measured resistance in two different measurement configurations, which are obtained by rotating the contact configuration 90º. The values of R_A and R_B are obtained from averaging the resistance values of four equivalent configurations, in order to account for non-uniformities on the sample geometry [53]:

$$R_A = \frac{R_{ij,kl} + R_{ji,lk} + R_{kl,ij} + R_{lk,jl}}{4} \quad (2.5)$$

$$R_B = \frac{R_{jk,li} + R_{kj,il} + R_{lj,ik} + R_{jl,ki}}{4} \quad (2.6)$$

There are up to 8 different configurations, as per each 90º rotation, current can be injected from two different contacts. Each configuration is calculated as follows:

$$R_{ij.kl} = \frac{V_{ij}}{I_{kl}} \quad (2.7)$$

Equation 2.7 defines the value of each $R_{ij,kl}$ value as a function of the voltage measured between the two neighbouring contacts i and j when a current is injected between the contacts k and l, as shown in Fig. 2.5. Ideally, both A and B configurations should lead to the same resistance value in a perfect square-shaped sample, which would result in f(Q) = 1 (from Eq. 2.4). For the rest of geometries, f(Q) < 1.

With respect to Hall effect measurements, the Hall coefficient can be calculated as [53]:

$$R_{H,s} = \frac{R_H}{t} = \frac{1}{B}\frac{R_{ik,jl} + R_{jl,ik}}{2} \quad (2.8)$$

where $R_{H,s}$ is the Hall sheet coefficient, R_H the Hall coefficient and B the magnetic flux density, applied perpendicularly to the surface of the lamellae. The sign of the Hall coefficient contains information about the sign of the majority carriers in the sample. We define the values of $R_{ik,jl}$ in the Hall effect, similarly as in the van der Pauw configuration., as:

$$R_{ik,jl} = \frac{V_{ik}}{I_{jl}} \quad (2.9)$$

In Eq. 2.9, V_{ik} is also named the Hall voltage, whose sign contains information about the type of the majority carrier on the sample. The number of different possibilities in the Hall effect configuration are doubled, as the magnetic field can be oriented in two opposite directions, always perpendicular to the plane of the sample. Therefore, up to 16 different configurations are possible. First, the Hall voltage is obtained from both measurements having the applied magnetic field in opposite directions.

$$V_{ik} = V_{ik,+} - V_{ik,-} \tag{2.10}$$

where the sub-indices + and - denote the voltages measured at positive and negative magnetic field directions. By applying Eq. 2.9, the voltages and currents are converted to resistance values. Then, the calculation is similar to the same averaging process of the resistivity configuration. The indices j, k, l, and i are set so the current is always injected in opposite contacts, measuring the voltage in the remaining two (see Fig. 2.5). From Hall measurements, two key parameters of the sample, as sheet concentration n_s and carrier type and Hall mobility $\mu_{H,s}$ can be obtained:

$$n_s = -\frac{r}{qR_{H,s}} \tag{2.11}$$

$$\mu_{H,s} = \frac{R_{H,s}}{R_s} \tag{2.12}$$

where r is the Hall scattering factor, which depends on the temperature and the carrier concentration, and q is the charge of the carrier, including the sign of the carrier concentration. The Hall scattering factor, for Si, takes values close to the unity for both holes [55] and electrons [56]. During this thesis, $r \approx 1$ for all calculations. In both Eqs. 2.11 and 2.12 it was assumed that there was a majority carrier type, in which either electrons either holes were predominant over the other. In the case in which both types were present in similar concentrations, the Hall mobility and sheet concentrations derived from Eqs. 2.11 and 2.12 would be the effective mobility and carrier sheet concentration:

$$\mu_{s,eff} = \frac{\mu_p^2 p_s - \mu_n^2 n_s}{\mu_p p_s + \mu_n n_s} \tag{2.13}$$

$$n_{s,eff} = \frac{(\mu_p p_s + \mu_n n_s)^2}{\mu_p^2 p_s - \mu_n^2 n_s} \tag{2.14}$$

In Eqs. 2.13 and 2.14, μ_p and p_s are the mobility and sheet concentration of holes, respectively, while μ_n and n_s are the mobility and sheet concentration of electrons.

The experimental setup used in this thesis is composed of several equipment interconnected. The core of the cluster is the previously described Keithley SCS 4200, which contains several automatized protocols to measure both van der Pauw measurements, the resistivity and the Hall effect, in all configurations.

A. Janis CSC-450ST cryostat. This cryostat system uses liquid He in a closed cycle, together with a heating system to maintain a stable temperature in the range of 10-500 K in the sample holder. With the use of this equipment it is possible to study the different properties of our samples as a function of the temperature. The sample holder (Fig. 2.6), made from copper and covered in gold, is screwed

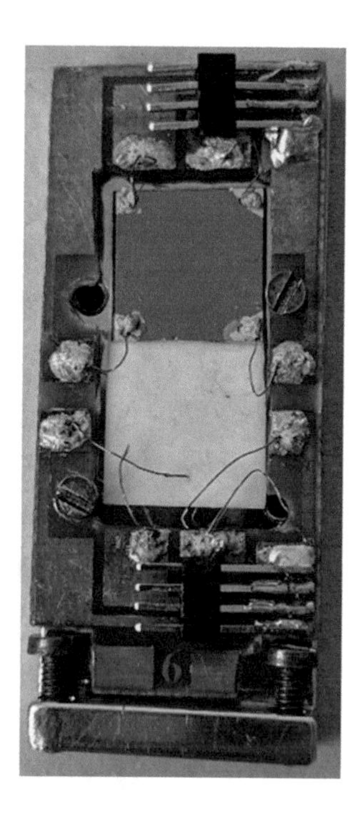

Fig. 2.6 Sample holder used for electro-optical characterisation inside the cryostat

to the cold finger of the cryostat.

B. Lakeshore 331 temperature controller.
C. Vacuum system. We used a turbomolecular pump, assisted by a membrane pump during this thesis to produce moderate to high vacuum levels inside the sample chamber of the cryostat. We kept the pressure in the order of 10^{-5} mbar to avoid condensation.
D. CTI Cryogenics 8200 Helium compressor. The cryostat needs liquid He to operate, which is provided by a closed-cycle He compressor (Fig. 2.7).
E. Alpha 7500 electromagnet. For Hall measurements, a magnetic field is necessary. The Alpha 7500 electromagnet is capable to produce magnetic fields up to 0.95 T.
F. KEPCO BOP 50-20 MG Bipolar current source. It is used to feed the electromagnet.
G. Lakeshore 455 DSP Gaussmeter.
H. Source and Measure Unit Keithley 2400. It is a programmable SMU used to feed electronic auxiliary equipment.

Finally, we have a lined-up four probe station, which uses the same physical phenomena as the van der Pauw configuration to obtain the sheet resistance of the sample. This set up is mostly used to map the resistivity of whole wafers, for material characterization mostly. The equipment used is a CMT-SR2000 manufactured by AITCO. We used a probe-to-probe separation of 1.016 mm with a tip radius of 40 μm.

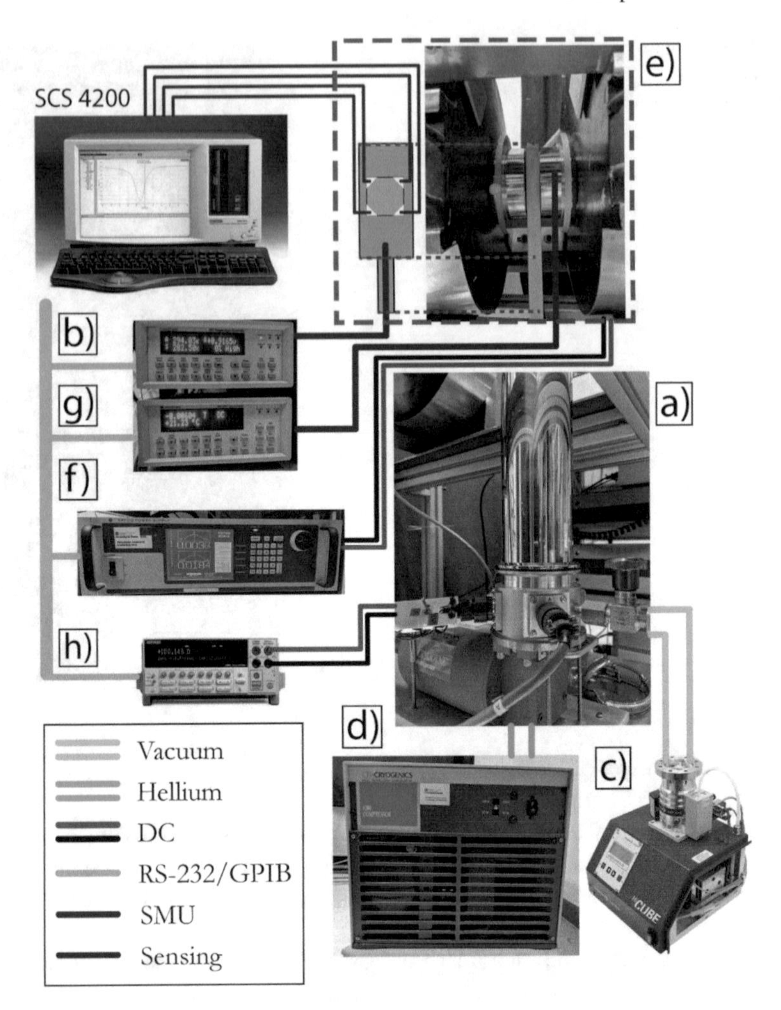

Fig. 2.7 Electrical characterisation cluster located in our laboratories

2.4.5 *Photoresponse*

One of the most important features of a photodetector is the change of its electrical properties under illumination, which we define as its photoresponse. The most common measurement on early stages of development of a photodetector is the photoresponse as a function of the wavelength of the impinging light. This characterisation process is especially relevant in supersaturated materials, as it is an indirect way to detect the presence of an intermediate or impurity band within the gap of the semiconductor [57]. This is usually done by illuminating the device with monochromatic light whose energy is lower than the bandgap.

The experimental setup is comprised of several instruments. First, a continuous light source is necessary. The light coming from the source is focused through an optical chopper, which blocks the path of the light in a periodic pattern, with tuneable frequency. The reason behind using chopped light is to enhance the sensitivity of the measurement using a lock-in amplifier [58]. After, light enters into the monochromator, which outputs a monochromatic beam, which is focused on the sample. As a result of the illumination, there might be a change on the electrical properties of the sample, which are sensed by the lock-in amplifier, synchronised to the same chopping frequency as light [58]. The measurement is recorded by a computer and later compared to a reference sensor, which provides information about the power spectral density. Depending on the device type the measurement process or the magnitude to be measured could be different. As it was described in Sect. 1.4 there are two main types of photonic detectors: photoconductive and photodiode detectors. In photoconductive detectors, we measure the photoconductance using a four-probe set up, as the van der Pauw configuration. A fixed value of the current is injected between two neighbouring contacts, while the resulting voltage is measured by the lock-in amplifier. All samples in this thesis are measured using 1 mA of current. The parameter of interest is the responsivity, defined as:

$$R_{sample}(\lambda) = \frac{V_{sample}(\lambda)}{P(\lambda)\cdot A_{sample}} \tag{2.15}$$

where V_{SAMPLE} is the bias produced by the sample measured using the lock-in amplifier, as a function of the wavelength of illumination, P(λ) is the optical incident power density and A_{SAMPLE} is the area of the sample. We measure the incident optical power P(λ) using a calibrated detector. The responsivity curve of the calibrated detector R_{SENSOR} is provided by the manufacturer. Therefore, we can obtain the optical incident power as:

$$P(\lambda) = \frac{V_{sensor}(\lambda)}{R_{sensor}(\lambda)\cdot A_{sensor}} \tag{2.16}$$

where V_{SENSOR} is the voltage produced by the sensor, measured using the lock-in amplifier as a function of the wavelength and A_{SENSOR} is the area of the sensor.

In the case of photovoltaic detectors, the contact configuration is different. They are usually contacted in both layers, the silicon substrate and the Ti implanted layer. Depending on the nature of the electrical junction between both materials, it is possible to measure the short-circuit current, the open-circuit voltage or the output current in any other polarisation between the electrodes. In this thesis, we will refer to this configuration as "transversal" characterisation.

In Fig. 2.8, the uppercase L stands for lock-in measurements, as it could be either a voltmeter or an ammeter, depending on the measurement configuration. In this thesis, we have predominantly measured spectral short-circuit photocurrent, as it allows a direct calculation of the External Quantum Efficiency (EQE), a key parameter in photovoltaic detectors and solar cells.

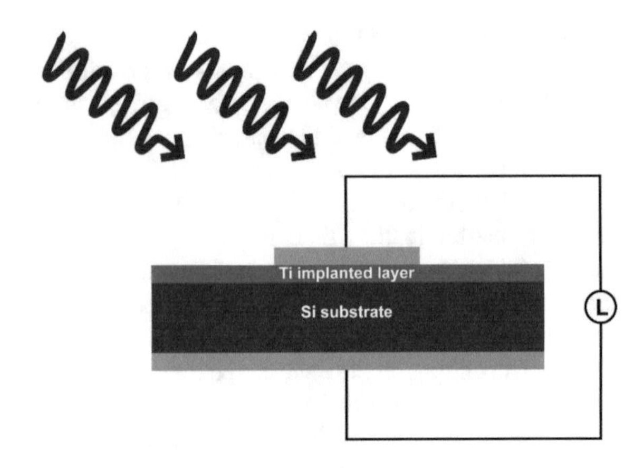

Fig. 2.8 Transversal photodiode characterisation with two electrodes, contacting the substrate and the implanted layer

$$EQE(\lambda) = \frac{carriersgenerated/s}{arrivingphotons/s} x100 \tag{2.17}$$

The numerator in Eq. 2.17 is obtained by dividing the short-circuit photocurrent to the charge of each carrier. The denominator, the number of arriving photons to the sample per second, is defined as the incident photon flux. In order to obtain the incident photon flux, we need to calculate the energy of each photon, using the well-known Planck's law. Later, we divide the incident optical power density (measured using the calibrated detector) to the energy of each photon. Finally, the incident photon flux density can be expressed as:

$$\phi_0(\lambda) = \frac{\lambda \cdot V_{sensor}(\lambda)}{q \cdot R(\lambda) \cdot hc \cdot A_{sensor}} \tag{2.18}$$

where q is the elemental charge, h is the Planck constant and c the speed of light in vacuum. ϕ_0 has units of $cm^{-2}s^{-1}$. After, the photon flux of each sample is obtained by multiplying the photon flux density by the area of the sample. Similarly, we could use the responsivity as described in Eq. 2.15. In the particular case of photovoltaic detectors, the responsivity can be expressed as a function of the current or the voltage per incident optical power, being their units A/W and V/W, respectively.

In our research laboratories, we have two different clusters related to spectral characterisation of the samples: the infrared cluster and the UV/Visible cluster. Most of the elements are common, but there are some differences which will be detailed in the following.

2.4.5.1 Infrared Photoresponse Cluster

Its purpose is to characterise samples in the infrared part of the spectra, from 1 to 13 μm at variable temperature, using the previously described Janis cryostat system for the temperature characterization. The cluster is composed of several instruments.

A. Light source. 140 W Newport 6363 SiC glow bar (Globar) source.

B. Thorlabs MC2000 optical chopper.
C. Bentham TMc300 IR monochromator. Range: 1 to 13 μm. Focal length: 300 mm.
D. Stanford SR-830 DSP dual phase lock-in amplifier.
E. Bentham 605 DC stabilised current source. Used to feed the Globar filament.
F. Keithley 2636A SMU. Already described in Sect. 2.4.3.
G. Janis CSC-450ST He cryostat. Already described in Sect. 2.4.4.
H. Bentham DH-PY pyroelectric detector. It is the detector used to calibrate the system. It is calibrated from 1 to 30 μm, at a fixed frequency of 15 Hz. If features a square-shaped sensor of 4 mm^2.
I. Homemade isolating box for micro-photovoltaic detectors.

2.4.5.2 Ultraviolet and Visible Photoresponse System

This cluster is used to characterise samples from 250 up to 1100 nm. It is complementary to the IR cluster; some of the instruments are shared. Unfortunately, the UV-Vis cluster is not yet compatible with the cryostat system. Therefore, all measurements in the UV-Vis range must be done at room temperature (Figs. 2.9, 2.10 and 2.11).

A. Light source. 250 W halogen type bulb with a cylindrical helix filament.
B. C-995 optical chopper manufactured by Terahertz Technologies Inc. (TTI).
C. Horiba iHR320 spectrometer. Range: 250 to 1100 nm. Focal length: 320 mm.
D. Stanford SR-830 DSP dual phase lock-in amplifier.
E. HP 6268B DC power supply. Used to polarise the halogen bulb.
F. Keithley 2636A SMU.
G. Homemade isolating box for micro-photovoltaic detectors.

Fig. 2.9 Homemade isolation box for micro-detectors, along with its coupling devices for the monochromators

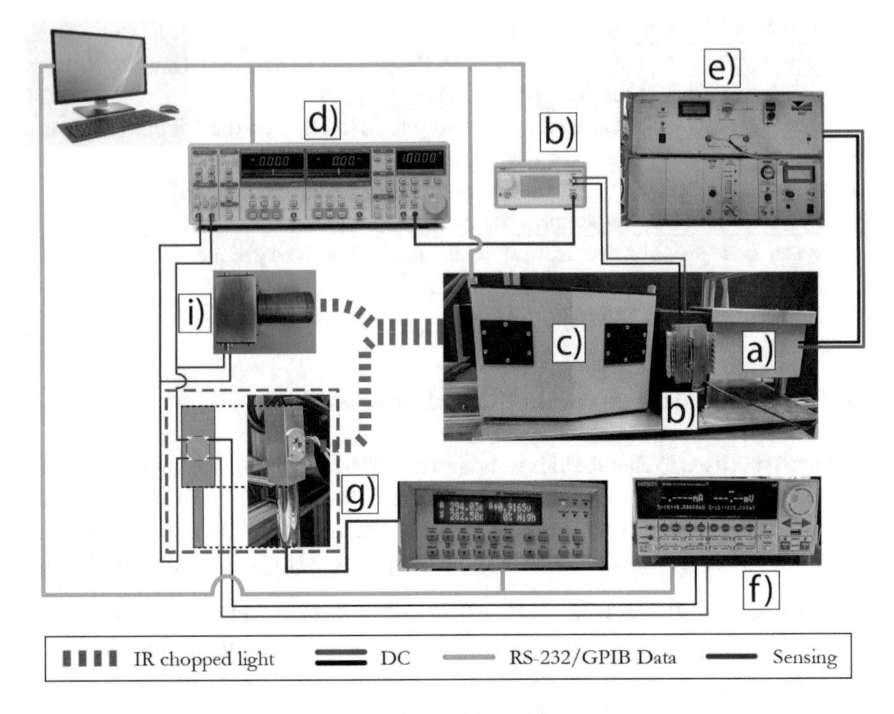

Fig. 2.10 IR photoresponse cluster located in our laboratories

Fig. 2.11 Homemade metallic box for resistivity and Hall effect measurements at room temperature

H. Homemade metallic box for photoconductors in van der Pauw configuration.
I. Hamamatsu S2281 Si calibrated photovoltaic detector. It is calibrated from 250 to 1180 nm. Detector area: 100 mm^2 (Fig. 2.12).

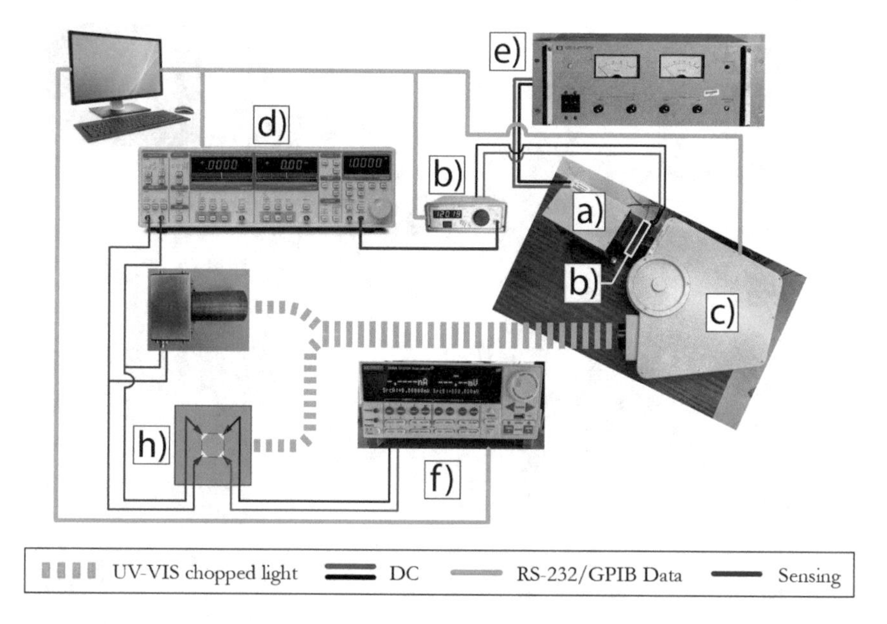

Fig. 2.12 UV-Vis photoresponse cluster located in our laboratories

2.5 Microelectronic Fabrication Techniques

The techniques described in this section are performed in the cleanroom facilities of the CAI Técnicas Físicas of the Faculty of Physics, with a total surface of 110 m^2.

2.5.1 *Photolithography*

Photolithography is used in semiconductor processing to selectively modify the properties of the sample on selected areas. A light-sensitive material, called photoresist, is used to cover the whole surface. The photoresist is exposed to UV light through a mask, which modifies the solubility of the exposed areas. With the use of a solvent, parts of the photoresist are dissolved, leaving a patterned surface. The uncovered portion of the surface may be later modified by means of other processes, as doping, etching or depositing techniques, mainly. By performing several photolithography processes, it is possible to fabricate devices, usually in the micro- or nanoscale.

In this thesis we have used a UV photolithography process to fabricate our devices. More details can be found in Sect. 3.5, where the fabrication route of the microscale devices is described.

The photolithography cluster is composed of several equipment.

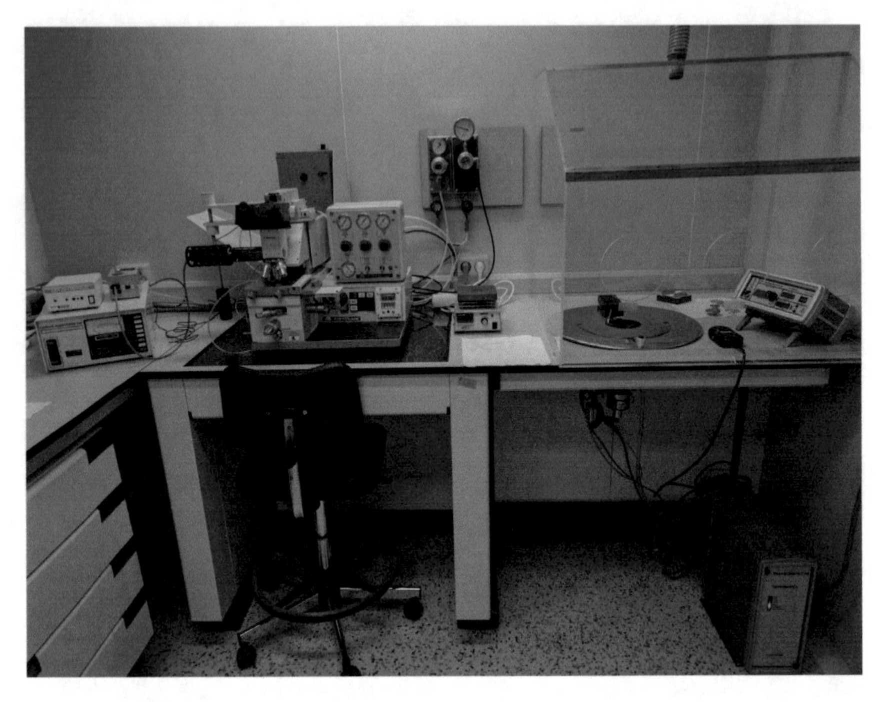

Fig. 2.13 Photolithography cluster. From left to right: lithography controller, mask aligner, hot plate furnace and spinner

A. MRC HP-1D hot plate furnace. It is used for the thermal treatments or "bakes" of the photoresist.
B. PWM101 Headway Research spinner. It is used to deposit thin layers of photoresist on the samples.
C. Karl-Suss MJB3 Mask aligner. It features a Hg lamp with emission lines at 405 nm (h-line) and 365 nm (i-line). It currently has installed a mask holder that allows up to 2" wafers.
D. Süss-Microtech CIC1200 lithography controller. It features a stabilised current source to feed the Hg lamp. It also controls the irradiance at the two emission lines (Fig. 2.13).

2.5.2 *Electron Beam Evaporation*

Electron beam (e-beam) evaporation is a highly anisotropic deposition technique, in which an electron gun directs electrons inside a refractory vase (crucible) containing the material to be deposited [59]. The material is heated up until it starts to release atoms or molecules, which are eventually deposited onto the sample.

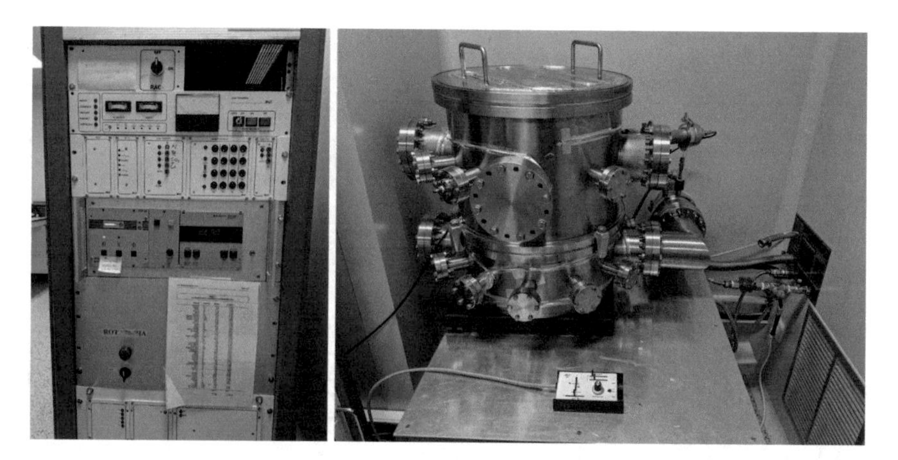

Fig. 2.14 Homemade e-beam evaporator. Left: controller rack. Right: vacuum chamber

The vacuum chamber of our experimental system is homemade, while the electronic equipment is commercial. The system works in high vacuum, in the order of $10^{-6}-10^{-7}$mbar to avoid oxidation of the evaporated molecules or atoms, especially titanium atoms [60], as they are oxygen scavengers [61]. It can contain up to four different elements to evaporate: Al, Ti, Pt and SiO_2, mounted on an automated rotating tray, controllable from the outside by the user. This allows the possibility to deposit a stack without breaking the vacuum. We use e-beam evaporation processes in metallization and passivation processes (Fig. 2.14).

2.5.3 *Electron Cyclotron Resonance-Chemical Vapour Deposition*

Electron Cyclotron Resonance-Chemical Vapour Deposition (ECR-CVD) is a deposition technique, where electrons are accelerated in a spiral trajectory by an alternating electric field, superimposed to a static magnetic field. The resonating chamber is filled with one or more gaseous species, which may be ionized by the electric field, releasing the aforementioned electrons. The gases may interact between them, producing stable atoms or molecules that may be adsorbed by the walls and the sample holder [62] (Fig. 2.15).

We use an Astek ECR 3 model system to deposit passivating layers, mainly silicon dioxide, silicon nitride or un-doped amorphous silicon. In the case of silicon oxide or silicon nitride, the same passivating layer may act as an Anti-Reflecting Coating (ARC) layer [63]. The source gas for silicon is a mixture of 5% silane (SiH_4) in 95% N_2.

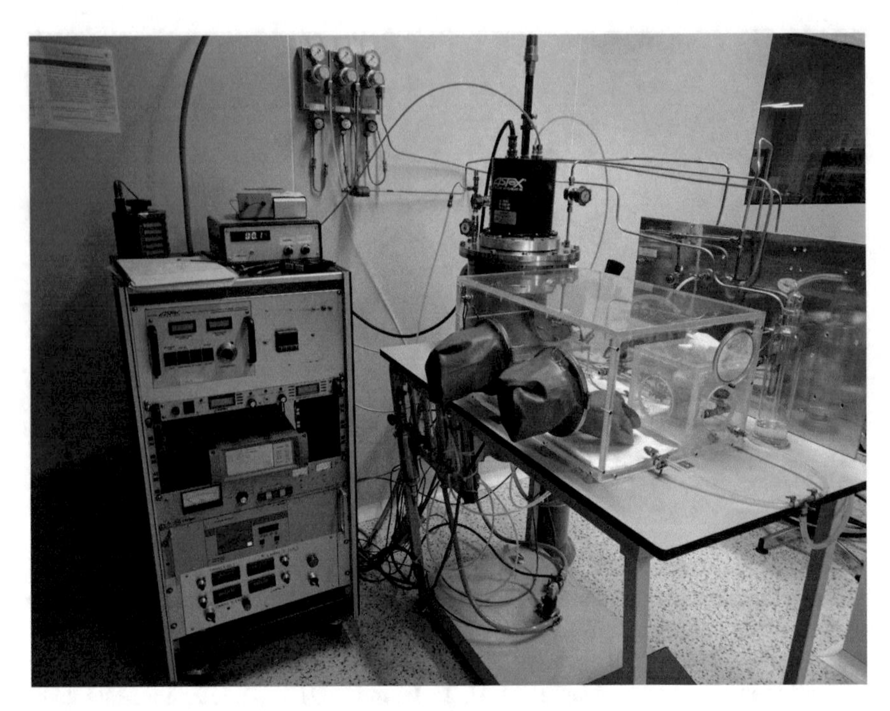

Fig. 2.15 ECR-CVD deposition equipment located in the cleanroom

2.5.4 Sputtering

Sputtering is a deposition technique that uses ionised gases to sputter a solid target, by using RadioFrequency (RF, usually at 13.56 MHz) or DC generators. The ions bombard the target, releasing atoms from it, which are later deposited on the walls of the vacuum chamber and the substrate holder.

A sputtering system for depositing metallic layers has been implemented during this thesis, due to the need to improve the contact quality in the fabricated devices. It has been completely designed and mounted by my thesis tutor and I for this thesis. It took us several months of design plus assembling. The idea was to design a sputtering system capable to hold up to three different metallic targets to deposit stacks in the same process. This restriction imposed a rotating sample holder, which could face any of the cathodes for the deposition process, while being RF compatible, with its own water cooling mechanism. This restriction made necessary a design able to withstand the high vacuum levels present inside the chamber, while minimizing water and vacuum leakages when rotating.

Due to the need to obtain ohmic contacts for device fabrication without further thermal processes, we aimed to fabricate a substrate holder compatible with RF as well to sputter the sample. That way, it would be possible to eliminate the native oxide, leaving an exposed bare silicon surface where the metals would be later deposited, all in the same machine, without breaking the vacuum. The three cathodes are necessary to deposit metallic stacks that could guarantee a good contact in both p-type and n-type silicon, using different metals for each type.

All pieces were fabricated in the workshop of the faculty, the CAI Taller Mecánico de Asistencia a la Investigación. More than 100 pieces were designed and fabricated from different materials: paramagnetic stainless steel, copper, brass and Teflon (Figs. 2.16, 2.17 and 2.18).

In Fig. 2.19 left, top cathode is tungsten (W), rear cathode is titanium (Ti), front cathode is aluminium (Al), the sample holder is on the right, the window and gas inlets are on the left and the vacuum valve on the lower flange. In our experimental set

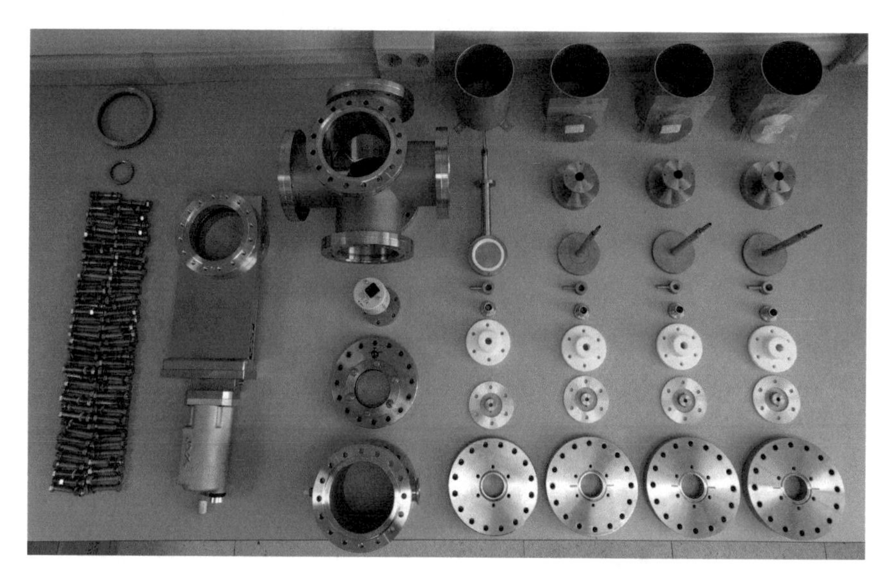

Fig. 2.16 Homemade sputtering system before assembling

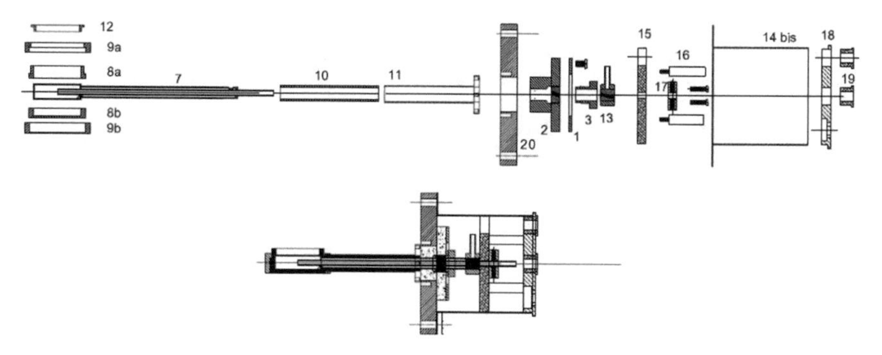

Fig. 2.17 Sample holder schematic in cross-section. Upper figure: exploded view. Lower figure: assembled

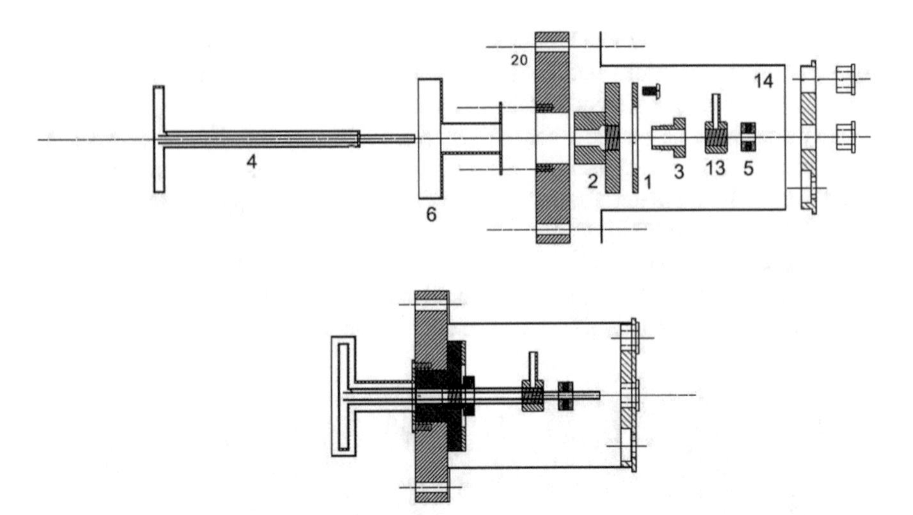

Fig. 2.18 Cathode holder schematic in cross-section. Upper figure: exploded view. Lower figure: assembled

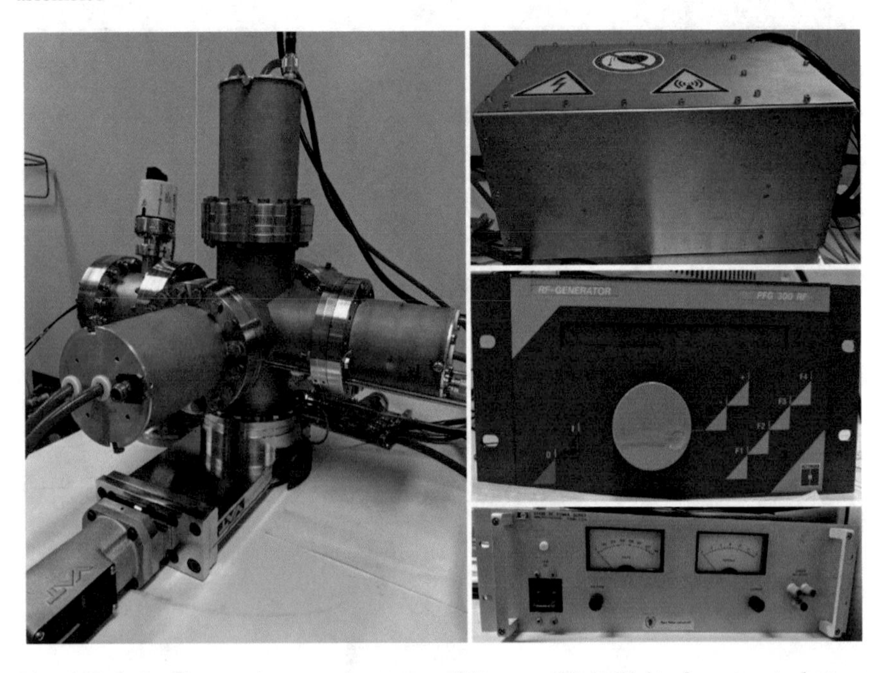

Fig. 2.19 Left: Homemade sputtering system fully assembled. Right: from top to bottom: matchbox, RF source and DC source

up, we use a Hüttinger PFG 300 RF generator and a matchbox Hüttinger PFM 1500 A. We have also a HP 6448B DC current source to perform future DC deposition processes. We have chosen Ar as the main gas to perform all sputtering processes. It is the most widely used gas for sputtering processes and it does not chemically react with any of the cathodes.

2.5.5 *Dry Etching*

Reactive Ion Etching (RIE), is a specific dry etching technique that uses ionised atoms, which chemically react with the surface of the sample. There is also an inherent sputtering (physical interaction) effect of the technique. Hence, the etching process using RIE is driven by two different mechanisms: an anisotropic one related to the RF power, where atoms are etched by physical bombardment and an isotropic one, related to gas pressure, where substrate atoms are etched by a chemical reaction with the ionised gas. For silicon etching the chosen gas is usually SF_6.

In this thesis, we use RIE to etch trenches, which may isolate pixels from each other in a matrix [64]. The experimental set up consists of a homemade vacuum chamber, containing a sample holder, compatible with up to 50.8 mm diameter wafers (Fig. 2.20).

In our experimental set up, we use a Hüttinger PFG 300 RF generator and a Hüttinger PFM 1500 A matchbox. The turbomolecular pump controller is a Pfeiffer vacuum DCU module.

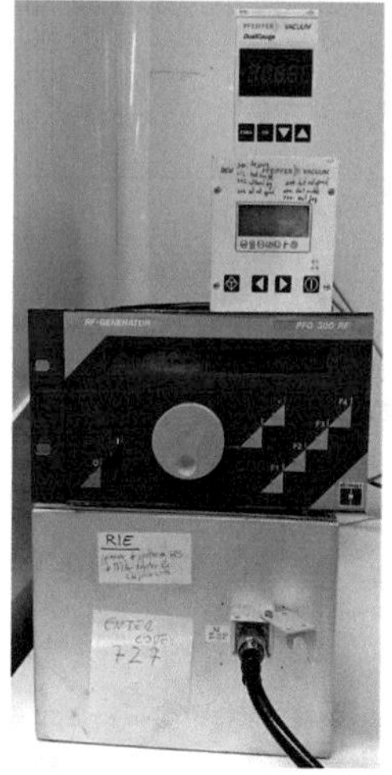

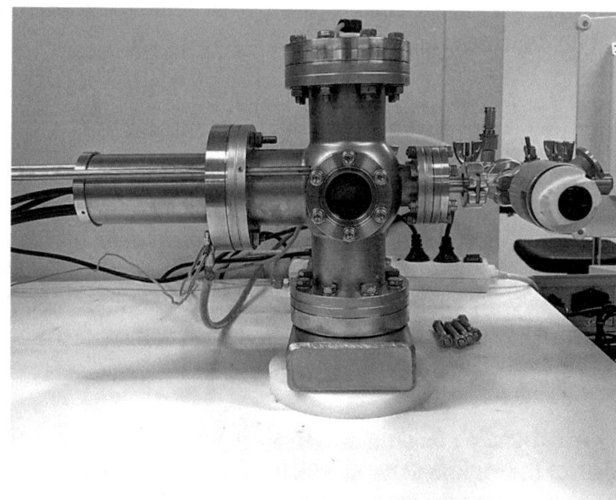

Fig. 2.20 Homemade Reactive Ion Etching equipment. Left: from top to bottom: vacuum sensor, turbomolecular pump controller, RF generator and matchbox. Right: RIE vacuum chamber

Table 2.3 Wet etching processes used in this thesis

Material	Formula	Etchant	Etchant solvent
Silicon dioxide	SiO_2	1:50 HF: H_2O	DI water
Silicon nitride	Si_xN_y	1:10 HF: H_2O	DI water
Aluminium	Al	1 g NaOH + 100 ml H_2O	DI water
Titanium	Ti	1:1:20 HF: H_2O_2: H_2O	DI water
Gold	Au	40 g + 1 g + 40 ml KI: I_2: H_2O	DI water
Positive PR (unexposed) Negative PR (exposed)	–	Developer	DI water
Positive PF (exposed) Negative PR (unexposed)	–	DiMethyl-SulfOxide (DMSO)	IPA
Silver paint	–	Acetone	IPA

2.5.6 *Wet Etching*

Wet etching is a common term which translates to chemical etching by means of liquid-to-solid reactions, being the solid the semiconductor sample and the liquid the chemical etchant. There is a wide variety of wet etching processes, as much as materials used in the microelectronics industry. We will also consider the chemical photoresist removal step as wet etching process. The etching processes performed in this thesis are indicated below.

In Table 2.3, PR stands for PhotoResist, while DI water stands for DeIonised water. The total time of reaction is strongly dependent on several factors, as temperature and the thickness of the deposited layer to etch.

2.5.7 *Wire Bonding*

Wire bonding is the technique that allows an efficient interconnection of nano- and microscale devices to a manoeuvrable macroscale chip carrier. A chip carrier is a piece made of various materials (plastic, ceramic, metallic) with metallic tracks. The tracks carry the information from the semiconductor chip towards the rest of the electrical circuit, in which the chip carrier is embedded (other electronic components or measurement equipment).

Wire bonding technique consists in welding a metallic wire or ribbon to a metallic track of the semiconductor, which is later welded to the track of the chip carrier. Since the track size in the semiconductor tends to be as small as possible to reduce the size of the chip, the wire diameter or ribbon width must be in the micrometre scale. Gold is the most commonly used metal, due to its ductility and malleability, while being particularly chemically stable, as it is less affected by corrosion or oxidation. The wire is not melted by applying just heat to the wire. The wire is guided using a tip, which vibrates, transmitting the energy to the wire. The tip also applies a slight force towards the sample. The combination of both processes softens the wire during a fraction of second, long enough to produce what is called an "ultrasonic bond" [65]. Wire bonding, by means of ultrasonic deformation, is almost the unique wire bonding method used at both industry and research levels. There are two main wire bonding techniques: ball bonding and wedge bonding. The first one uses a needle with a single cavity that goes in the symmetry axis. The wire comes out through the point of the needle. A ball is formed by melting the wire tip using a high voltage source. The second technique uses a flat edged needle with two different orifices and does not require the formation of a ball on the tip (Fig. 2.21).

Our experimental set up is formed by a semi-automatic wire bonder TPT HB16, capable of performing both ball and wedge bond welds with their respective tools. The TPT HB-16 is compatible with gold and aluminium wires with a diameter ranging 15 to 75 μm and gold ribbons up to 25 × 250 μm (different tools may be necessary). The ultrasound energy is provided by a piezoelectric transducer, with an oscillating frequency of 63.3 kHz (Fig. 2.22).

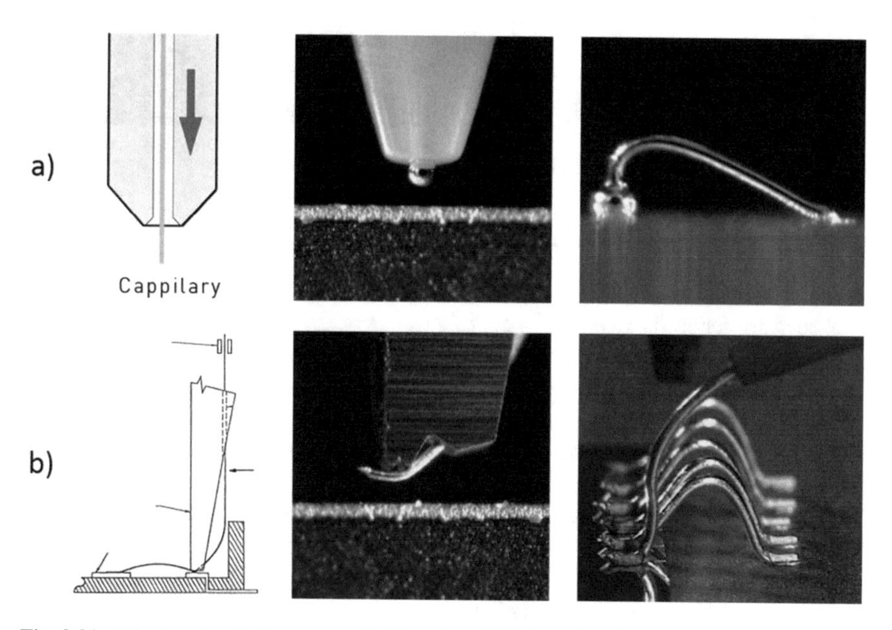

Fig. 2.21 Wire bonding techniques. **a**) Ball bonding. **b**) Wedge bonding (*Source* TPT HB16 Wire Bonder user manual)

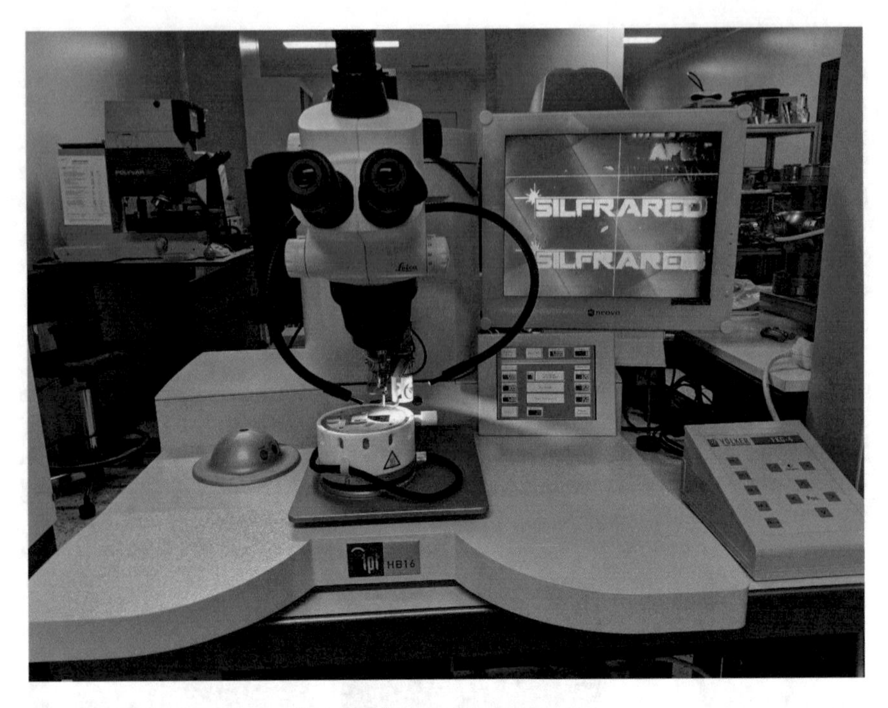

Fig. 2.22 TPT HB16 wire bonder located in our cleanroom

2.6 Pixel Architecture Characterization

This section generally describes part of the characterisation process of the pixel architecture used during my internship in STMicroelectronics in Crolles (France). The starting wafer size is 300 mm diameter, while devices are based on CMOS Image Sensor (CIS) technology. All the information related to these samples is contained in Chapter 5, although most of the details are protected by a Non-Disclosure Agreement (NDA).

The test chip we used (also labelled die), contains, rounding up the numbers, 10^6 pixels. During the characterisation process, each pixel can be monitored individually. For each measurement, we are analysing the behaviour of a population of pixels, which means that we must apply statistical models to analyse the amount of available data.

There are two different ways to measure the electrical and optical properties of the fabricated dies. In a first approach, devices are measured straight from the wafer, by using instruments featuring point tools, directly contacting the semiconductor, which is called Wafer Level Characterisation (WLC). These processes are fast and allow the mass characterisation of full wafers. The wafers follow a standard measurement procedure and there is little room for improvisation or further researching. The second way of characterisation is at the die level. Wafers are cut into dies, which are manually measured by the operator, which has the freedom to choose the appropriate test and measurement conditions. We have performed both characterisation possibilities.

2.6.1 *Pixel Characterisation*

The interest on pixel characterisation parameters and techniques is growing in the last years. Several of them are standard in the industry, while others are specific of each technology. The parameters described here are common in CMOS image sensors, especially for Pinned-PhotoDiode (PPD) CIS and similar technology. Most common ones are briefly described here, in the usual order of calculation, as some parameters depend on others. More information on CIS characterisation can be found elsewhere [66–68].

To perform the measurements, the die is attached to a motherboard, which acts as a power supply and a communication interface with a computer. The motherboard, specifically designed for the pixel architecture under study, reads in real time the data of each pixel, which translate the numerical value into brightness. The software that controls the motherboard also offers several options to filter and display information coming from different pixel configurations. The user may choose to obtain and represent the average value of any measurement of a chosen pixel population to perform statistical studies.

2.6.1.1 Photon Conversion Characteristic

The first measurement that is performed on each die is the Photon Conversion Curve (PCC). The die is illuminated with polychromatic light at a fixed intensity. The photons may produce a charge accumulation inside the pixel, which is converted to voltage by the ROIC. The output voltage (labelled average swing) depends on both the illumination power and the amount of time that the pixel is collecting charges, the integration time. The PCC is the result of plotting the average swing as a function of the integration time.

There are two main regions: the low swing range, which is usually linearly increasing with the integration time, and the saturation range, where the average swing reaches a maximum (saturation) value. The linear region is important because it is the range in which the pixel is supposed to work. From the saturation regime we obtain the saturation voltage, a key value that is later used for other calculations (Fig. 2.23).

2.6.1.2 Noise

Noise in a CMOS Image Sensor is intended as a spurious signal that affects the system, degrading the signal quality. There are two noise classifications, attending

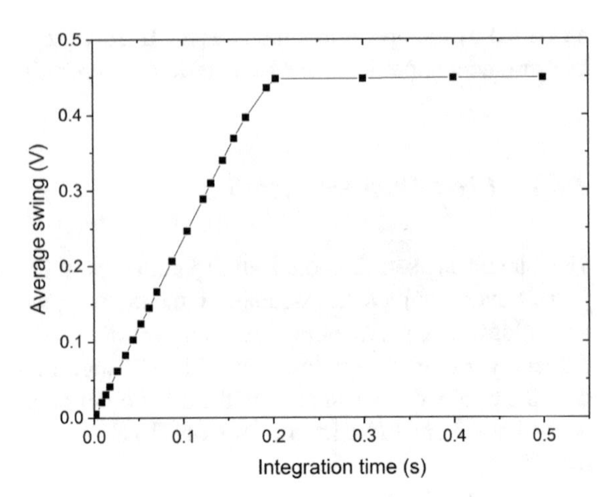

Fig. 2.23 Photon conversion curve of a CMOS image sensor

to their nature: temporal and spatial noise. Temporal noise is related to the natural variation of any noise source over time inside the same pixel. It is usually described using statistical models, and can be minimised by doing several measurements and averaging them later. On the opposite, spatial noise is taken into account when examining the average noise of a group of pixels. Each pixel may have different noise levels according to slight changes in the fabrication processes. The spatial noise is also called pattern noise, which is divided into Fixed Pattern Noise (FPN), which accounts for non-uniformities of the lecture chain (row or column dependant) at dark, and Photon Response Non-Uniformity (PRNU), which accounts for the pixel non-uniformity under illumination. PRNU may include different light response of the filters or microlenses above the pixel itself. Of course, each individual pixel of this group still has its own fluctuating temporal noise [69]. In general, pattern noise is die dependent. Each die may have different FPN or PRNU as their fabrication process may be slightly different. It can be minimised studying the response of each pixel at uniform light and at dark, as pattern noise does not change significantly between different measurements. The processing software is able to detect the non-uniformities and apply correction factors to each pixel, column or row before displaying the final image. Basically, every link in the photon-to-signal conversion chain has its related noise. In general:

$$Noise_{total} = \sqrt{\sum noise_i^2} \tag{2.19}$$

The first link of the chain is the noise associated to photons themselves. The number of photons arriving to a detector during the exposure time follows a Poisson distribution [69]. It can be demonstrated that the average photon shot noise follows the next relationship [70]:

$$\sigma_{ph} = \sqrt{N_{ph}} \tag{2.20}$$

where σ_{ph} is the photon shot noise and N_{ph} is the average photon count value. Photon shot noise is an unavoidable noise source in any photoresponse measurement of any device. Photon noise depends on the square root of the number of photons arriving to the detector. It implies that photon noise will be dominant in low light conditions.

The next link in the noise chain would be the noise coming from the active area of the pixel. Photons may have been absorbed, possibly forming electron-hole pairs (which is the purpose of the photodetector) which must travel from the generation place towards the lecture chain. The semiconductor may also have intrinsic noise, coming from the dark current, which will be assessed later. The transistors involved in the lecture chain may also introduce noise, usually referred to as "Read noise" [71]. In overall, most noise sources can be minimised using several strategies during the fabrication or the measurement process, except the Photon Shot Noise, which will play a key role in the characterisation process of the pixel.

2.6.1.3 Dark Current

Dark current can be obtained from the following relation:

$$I_{dark} = \frac{q_{dark}}{T_{int}} \tag{2.21}$$

It expresses the amount of stored charges q_{dark} during the integration time T_{int}, and it is often measured in amperes or in electrons per second. Dark current is the term that refers to the remaining charges inside the pixel at dark, after the minimisation of other noise sources, as pattern noise, mainly. As it is measured in dark conditions, photon shot noise is not present. The remaining contributions to noise usually come from the thermal carrier generation inside the semiconductor, or generation processes mediated by surface or bulk defects. Within bulk defects, most of them are produced by non-annealed defects coming from the ion implantation process or due to contamination [72].

2.6.1.4 Photon Transfer Curve

The Photon Transfer Curve (PTC) is the curve obtained when noise is represented as a function of the signal in a double log scale [73].

Figure 2.24 depicts the typical PTC of a CMOS image sensor. Depending on which noise source is predominant, the shape of the PTC may change. In a virtually noise-free CMOS detector, Read noise and FPN regimes would not be present, leading to a PTC with a constant slope of 0.5, starting from zero, up to the saturation regime, where the curve collapses due to carrier overflow inside the pixel [73]. FPN has been found to be linearly dependent on the signal, thus showing a slope close to unity [73]. The PTC is useful to find out the regime in which the noise of the pixel array is dominated by the photon shot noise, which allows the calculation of the Conversion Voltage Factor, another key parameter of the CIS.

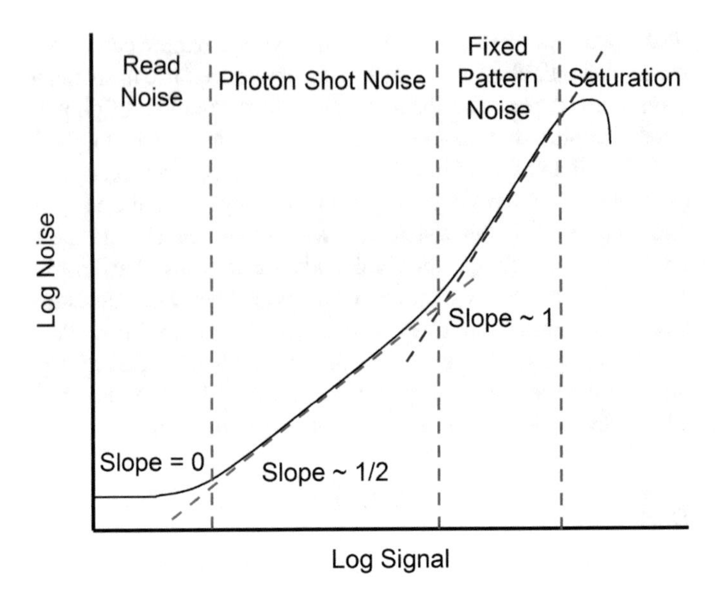

Fig. 2.24 Graphic example of a Photon Transfer Curve of a CMOS Image Sensor

2.6.1.5 Conversion Voltage Factor

The Conversion Voltage Factor (CVF) is a key parameter widely used in the pixel characterisation process. Its definition is [72]:

$$CVF = \frac{V_{pix}}{N_e} \tag{2.22}$$

That is, the CVF is the ratio of the measured voltage in the output of the pixel V_{pix} per each electron collected N_e. It is often measured in $\mu V/e^-$, if collected carriers are electrons, or $\mu V/h^+$ in the case of holes. The importance of the CVF lies in the straight-forward calculation of the amount of carriers inside the pixel just by looking at the output voltage. The carrier count inside the pixel is crucial for determining one of the most important parameters of a pixel, which is its quantum efficiency. It is demonstrated that, if the main noise contribution source is the Shot photon noise, the CVF can be experimentally extracted as [72]:

$$CVF = \frac{\sigma_V^2}{V_{pix}} \tag{2.23}$$

where σ_V^2 is the variance of the voltage measured at the pixel. The graphical representation of the variance as a function of the signal has been also labelled as Photon Transfer Curve by other authors [72].

Two different regimes are observed in the PTC shown in Fig. 2.25: a first linear regime up to around 0.40 V, and the saturation regime for higher values. CVF can be extracted by doing a linear regression in the linear regime. In Fig. 2.25 we find a

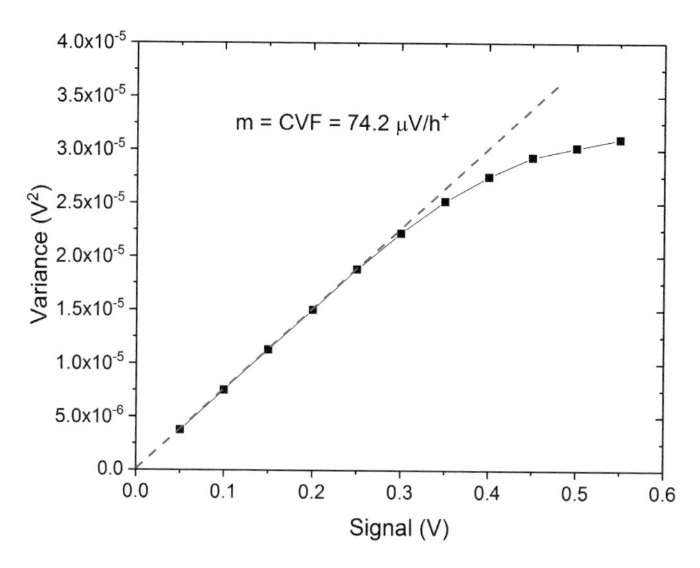

Fig. 2.25 Example of Photon Transfer Curve measured during this thesis

slope of 74.2 μV/h$^+$, which corresponds to the CVF for this particular measurement. Values between 50-150 μV/h$^+$ are usually found on commercially available CMOS image sensors [72].

2.6.1.6 Signal-to-Noise Ratio

Signal-to-Noise Ratio (SNR) is the ratio between the signal and the noise at a given input level. It is expressed as:

$$SNR = 20 \log \frac{V_{pix}}{V_{noise}} \tag{2.24}$$

where higher SNR values are preferred. An analogous formulation would be to use the amount of charges instead of the voltage, through the CVF.

2.6.1.7 Full Well Capacity

Full Well Capacity (FWC) is the maximum number of carriers that a pixel can hold before saturating the pixel. Carriers are accumulated inside the pixel by a potential barrier (as seen in Fig. 1.6). As carriers accumulate inside the potential well, the potential barrier decreases, until the point of saturation, where the potential well disappears. At this point, the pixel will no longer accept more charges. The FWC is directly related to the saturation voltage, V_{sat}, and it is calculated just by translating the saturation voltage to the number of charges using the CVF:

$$FWC(e^-, h^+) = \frac{V_{sat}}{CVF} \quad (2.25)$$

It is desired to have the highest FWC as possible, as it is directly related to the maximum dynamic range of the pixel, the ratio of the highest illuminance that a pixel can measure over the lowest value of illuminance it can detect.

2.6.1.8 Photon Response Non-uniformity

Focal Plane Array CMOS Imaging Sensors are composed of large groups of pixels. As explained before, there is some expected variability in the response of each pixel due to the manufacturing process. Photon Response Non-Uniformity (PRNU) is part of the previously described pattern noise, a spatially distributed noise pattern within a group of pixels. Therefore, PRNU cannot be measured for a single pixel. It gives information of the relative standard deviation, in percentage, with respect to the averaged value of photoresponse. It is usually measured at red, green and blue colours, as the final version of the pixel would be covered by coloured filters. PRNU can be also measured in white light. Usually, PRNU is represented as a function of the fill factor of the pixel, which is the amount of collected carriers. Best conditions are when the PRNU is as low as possible for all hole collection values. PRNU value is important as it provides direct information on the pixel uniformity under illumination. Although PRNU can be minimised using ulterior data treatment, it is preferred to develop pixels as homogenous as possible. PRNU is commonly used to evaluate the homogeneity of the fabrication process.

In our study on Ti supersaturated Si layers, PRNU can be used as a powerful tool to evaluate the homogeneity of the implanted material, by comparing the PRNU of a reference die (non-implanted) and an Ti implanted die. Among the parameters which are important in CMOS industry, homogeneity is highlighted, especially when introducing a new material in a commercial device.

2.6.1.9 Quantum Efficiency

The quantum efficiency measures the percentage of light that is converted into charges as a function of the wavelength. The die is illuminated with monochromatic light and the pixel voltage is measured as a function of the wavelength. Then, the voltage is converted to current using the CVF and the integration time. Since the incident optical power on the die is calibrated, we can obtain the number of photons incident per cm^2, similarly to how it was described in Sect. 2.4.5. Finally, QE measured on the dies can be calculated as:

$$QE(\lambda) = \frac{V_{out}(\lambda) \cdot hc}{\lambda \cdot CVF \cdot P_{in}(\lambda) \cdot A \cdot T_{int}} x100 \quad (2.26)$$

where V_{out} is the average swing at each wavelength, P_{in} is the incident power in W/cm^2, A is the sensor area in cm^2 and T_{int} in seconds the integration time. The higher the QE the better for device performance, as it could detect lower light conditions, requiring shorter integration times. Shorter integration times lead to lower noise values, as dark current is dependent on the integration time, which leads to better SNR values.

2.6.2 *Process Characterisation*

The manufacturing processes need to be monitored during and after the fabrication of the samples, to assure that they are under compliance. In this section, we describe only the extra steps we performed on the Ti ion implanted wafers to check the compatibility of the process during the last steps.

2.6.2.1 Current-Voltage Measurements

We measured IV curves of different pixel structures to test the effect of the Ti implantation and the laser annealing processes on the transistors used in the ROIC. We wanted to check that the ion implantation plus the laser process did not modify the behaviour of the Read and Source Follower transistors, which are located atop the pixel, in the opposite side of the ion implantation process (Fig. 1.5). We measured the drain current of both MOS transistors as a function of the drain-source voltage V_{DS}. On and off state current were analysed, as well as the threshold voltage V_T. We used the second derivative method of the I_D-V_{DS} curve and searched for the maximum value in the linear region of the transistor [74]. The V_{DS} value at which the maximum value of the second derivative is found is the threshold voltage.

$$V_T \approx \max \frac{\partial^2 I_D}{\partial V_{DS}^2} \tag{2.27}$$

2.6.2.2 Camera Acquisition Pictures

It is possible to use the own picture, taken by the CMOS image sensor in the optical characterisation bench, to obtain information about the pixel structure. To do that, the CMOS sensor is illuminated with uniform light. Using the picture acquisition software, we take a picture in light conditions, close to saturation. If there is some structural damage in the pixel, it may be visible in the image it provides. Although this damage effect may be visible in the statistics associated to the sensor, it may be helpful to compare the optical images obtained by the microscope and the pictures acquired by the sensor to see if there is a correlation.

References

1. Olea J (2009) *Procesos de implantación iónica para semiconductores de banda intermedia.* Thesis dissertation
2. Garcia-Hemme E (2015) *Respuesta infrarroja en silicio mediante implantación iónica de metales de transición.* Thesis dissertation
3. Shockley W (1954) *Forming semiconductive devices by ionic bombardment.* U.S. Patent 2,787,564
4. Collart EJH et al. (2010) *Process characterization of low temperature ion implantation using ribbon beam and spot beam On the AIBT iPulsar high current.* Ion Implantation Technology 2010 1321, 49
5. Dearnaley GK, Nelson RS (1969) *Ion implantation in semiconductors.* Phys Bull 20
6. Olea J, Toledano-Luque M, Pastor D, Gonzalez-Diaz G, Martil I (2008) *Titanium doped silicon layers with very high concentration.* J Appl Phys 104
7. Gibbons JF (1972) *Ion implantation in semiconductors 2. damage production and annealing.* Pr Inst Electr Elect 60:1062
8. Suzuki K, Kawamura K, Kikuchi Y, Kataoka Y (2006) *Compact model for amorphous layer thickness formed by ion implantation over wide ion implantation conditions.* Ieee Trans Electron Devices 53:1186–1192
9. Yang W et al. (2017) *Au-rich filamentary behavior and associated subband gap optical absorption in hyperdoped Si.* Phy Rev Mater 1
10. Liu F et al. (2018) *Structural and electrical properties of Se-hyperdoped Si via ion implantation and flash lamp annealing.* Nucl Instrum Meth B 424:52–55
11. Picraux ST (1969) *Channeling in semiconductors and its application to the study of ion implantation.* Thesis dissertation, 119
12. Zhang HP et al. (2008) *United Gauss-Pearson-IV distribution model of ions implanted into silicon.* Solid State Ionics 179:832–836
13. Olea J, del Prado A, Pastor D, Martil I, Gonzalez-Diaz G (2011) *Sub-bandgap absorption in Ti implanted Si over the Mott limit.* J Appl Phys 109
14. Garcia-Hemme E et al. (2013) *Far infrared photoconductivity in a silicon based material: Vanadium supersaturated silicon.* Appl Phys Lett 103
15. Ertekin E et al. (2012) *Insulator-to-metal transition in selenium-hyperdoped silicon: observation and origin.* Phys Rev Lett 108
16. Franta B et al. (2015) *Simultaneous high crystallinity and sub-bandgap optical absorptance in hyperdoped black silicon using nanosecond laser annealing.* J Appl Phys 118
17. Ting CY, Crowder BL (1982) *Electrical-properties of Al/Ti contact metallurgy for vlsi application.* J Electrochem Soc 129:2590–2594
18. Pankove JI, Lampert MA, Tarng ML (1978) *Hydrogenation and dehydrogenation of amorphous and crystalline silicon.* Appl Phys Lett 32:439–441
19. Miki T, Morita K, Sano N (1997) *Thermodynamic properties of titanium and iron in molten silicon.* Metall Mater Trans B 28:861–867
20. Cole JM, Humphreys P, Earwaker LG (1984) *A melting model for pulsed laser-heating of silicon.* Vacuum 34:871–878
21. Thompson MO et al. (1983) *Silicon melt, regrowth, and amorphization velocities during pulsed laser irradiation.* Phys Rev Lett 50:896–899
22. Deunamuno S, Fogarassy E (1989) *A thermal description of the melting of c-silicon and a-silicon under pulsed excimer lasers.* Appl Surf Sci 36:1–11
23. Green MA (2008) *Self-consistent optical parameters of intrinsic silicon at 300 K including temperature coefficients.* Sol Energ Mat Sol C 92:1305–1310
24. Mizoguchi HST, Itou N, Yamazaki T (2016) *Short wavelength light source for semiconductor manufacturing: challenge from excimer laser to LPP-EUV light source.* Komatsu Technical Report 62:11
25. Yashiro M et al. *Excimer laser gas usage reduction technology for semiconductor manufacturing.* Optical Microlithography XXX 10147

26. Basting D, Stamm U (2001) *The development of excimer laser technology—history and future prospects.* Z Phys Chem 215:1575–1599
27. Akane TNT, Matumoto S (1992) *Two-step doping using excimer laser in boron doping of silicon.* Jpn J Appl Phys 31:4
28. Venturini J et al. (2004) *Excimer laser thermal processing of ultra-shallow junction: laser pulse duration.* Thin Solid Films 453:145–149
29. Miyasaka M, Stoemenos J (1999) *Excimer laser annealing of amorphous and solid-phase-crystallized silicon films.* J Appl Phys 86:5556–5565
30. Reitano R, Smith PM, Aziz MJ (1994) *Solute trapping of group-iii, iv, and v elements in silicon by an aperiodic stepwise growth-mechanism.* J Appl Phys 76:1518–1529
31. Tang K, Ovrelid EJ, Tranell G, Tangstad M (2009) *Critical assessment of the impurity diffusivities in solid and Liquid silicon.* Jom-Us 61:49–55
32. Kuryliw EA (2003) Analyzing the thermal annealing behavior of laser thermal processed silicon. Thesis dissertation
33. Maszara WP, Rozgonyi GA (1986) *Kinetics of damage production in silicon during self-implantation.* J Appl Phys 60:2310–2315
34. Thompson MO et al. (1984) *Melting temperature and explosive crystallization of amorphous-silicon during pulsed laser Irradiation.* Phys Rev Lett 52:2360–2363
35. Ishihara R, Yeh WC, Hattori T, Matsumura M (1995) *Effects of light-pulse duration on excimer-laser crystallization characteristics of silicon thin-films.* Jpn J Appl Phys 1(34):1759–1764
36. Huet KBC, Negru R, Aing P, Venturini J (2010) *Full device exposure laser thermal annealing: high performance and high yield junction formation process.* 2010 18th international conference on Advanced thermal processing of Semiconductors (RTP), 4
37. Gyulai JPF, Krafcsik I, Solyom A, Riedl P, Bori L (1989) *Calibration of SIMS measurements by ion implantation.* Periodica Polytechnica 34:6
38. Russ JC (1984) *Fundamentals of energy Dispersive x-ray analysis.* Butterworths-Heinemann, 314
39. Vickerman JC, Briggs D (2013). *ToF-SIMS: materials analysis by mass spectrometry.* IM publications
40. Pastor D et al. (2011) *UV and visible Raman scattering of ultraheavily Ti implanted Si layers for intermediate band formation.* Semicond Sci Tech 26
41. Goldstein J, Newbury DE, Joy DC, Lyman CE, Echlin P, Lifshin E, Sawyer L, Michael JR (2003) *Scanning electron microscopy and x-ray microanalysis.* Springer US
42. Giannuzzi LA, Stevie FA (1999) *A review of focused ion beam milling techniques for TEM specimen preparation.* Micron 30:197–204
43. Binnig G, Quate CF, Gerber C (1986) *Atomic force microscope.* Phys Rev Lett 56:930–933
44. Toulemonde M et al. (1985) *Time-resolved reflectivity and melting depth measurements using pulsed ruby-laser on silicon.* Appl Phys a-Mater 36:31–36
45. Galvin GJ et al. (1983) *Time-resolved conductance and reflectance measurements of Silicon during pulsed-laser annealing.* Phys Rev B 27:1079–1087
46. Kerdiles S et al. (2016) *Dopant activation and crystal recovery in arsenic-implanted ultra-thin silicon-on-insulator structures using 308 nm nanosecond laser annealing.* 2016 16th international workshop on junction technology (IWJT), 72–75
47. Holsteyns FRJ, Toan Le Q, Kenis K, Mertens PW (2004) *Seeing through the haze: process monitoring and qualification using comprehensive surface data.* Yield Management Solutions KLA Tencor Magazine Spring 2004, 5
48. Halioua M, Liu H-CJO (1989) Engineering, l. i. *Optical three-dimensional sensing by phase measuring profilometry* 11, 185–215
49. Maley N (1992) *Critical investigation of the infrared-transmission-data analysis of hydrogenated amorphous-silicon alloys.* Phys Rev B 46:2078–2085
50. Liu XG et al. (2014) *Black silicon: fabrication methods, properties and solar energy applications.* Energ Environ Sci 7:3223–3263
51. Cobet C (2014) *Ellipsometry: a survey of concept.* Ellipsometry of functional organic surfaces and films 1–26, Springer

52. Neamen DA (1997). *Semiconductor physics and devices.* Vol. 3. McGraw-Hill, New York
53. Van der Pauw LJ (1958) *A method of measuring the resistivity and hall coefficient on lamellae of arbitrary shape.* Phillips Tech Rev 20:5
54. Gonzalez-Diaz G et al. (2017) *A robust method to determine the contact resistance using the van der Pauw set up.* Measurement 98:151–158
55. Lin JFLSS, Linares LC, Teng KW (1981) *Theoretical analysis of hall factor and hall mobility in p-type silicon.* Solid State Electron 24:6
56. Kirnas IGKPM, Litovchenko PG, Lutsyak VS, Nitsovich VM (1974) *Concentration dependence of the hall factor in n-type silicon.* Phys Status Solidi A 23:5
57. Garcia-Hemme E et al. (2014) *Room-temperature operation of a titanium supersaturated silicon-based infrared photodetector.* Appl Phys Lett 104
58. Scofield JH (1994) *Frequency-domain description of a lock-in amplifier* 62:129–133
59. Singh J, Wolfe DE (2005) *Review nano and macro-structured component fabrication by electron beam-physical vapor deposition (EB-PVD).* 40:1–26
60. Mattox DM (2010) *Handbook of physical vapor deposition (PVD) processing (Second Edition)* (ed Donald M. Mattox) pp 195–235. William Andrew Publishing
61. Gambino JP, Colgan EG (1998) *Silicides and ohmic contacts.* Mater Chem Phys 52:99–146
62. Lassig SE, Tucker JD (1995) *Intermetal dielectric deposition by electron cyclotron resonance chemical vapor deposition (ECR CVD).* Microelectron J 26, XI–XXIII
63. El amrani A et al. (2008) *Silicon nitride film for solar cells.* Renew Energy 33:2289–2293
64. Arnaud T et al. (2011) *Pixel-to-pixel isolation by deep trench technology: application to CMOS image sensor.* IISW Conference at Hokkaido, Japan
65. Harman GG (2010) *Wire bonding in microelectronics 3rd edition.* McGraw-Hill Education, 446
66. Ge X, Mamdy B, Theuwissen A (2016) *A comparative noise analysis and measurement for n-type and p-type pixels with CMS technique.* Electronic Imaging, Image Sensors and Imaging Systems 2016, pp. 1–6(6)
67. Goiffon V et al. (2014) *Pixel level characterization of pinned photodiode and transfer gate physical Parameters in CMOS image sensors.* Ieee J Electron Devi 2:65–76
68. Wang F, Theuwissen AJP (2019) *Pixel optimizations and digital calibration methods of a CMOS image sensor Targeting high linearity.* IEEE Trans Circuits Syst I: Regul Pap 66:930–940
69. Theuwissen A (2007) *CMOS image sensors: State-of-the-art and future perspectives.* Proc Eur S-State Dev, 21- +
70. Janesick JR (2007) *Photon transfer.* Spie Press Book
71. Hornsey RI (2008) *Noise in image sensors.* University of Waterloo Prints
72. Mamdy B (2016) *Nouvelle architecture de pixel CMOS éclairé par la face arrière, intégrant une photodiode à collection de trous et une chaine de lecture PMOS pour capteurs d'image en environement ionisant.* Thesis dissertation, 161
73. Gardner D *Characterizing digital cameras with the photon transfer curve.* https://pdfs.semanticscholar.org/8596/00415df0290714f8d928f40889f4eb6db5a2.pdf
74. Ortiz-Conde A et al. (2002) *A review of recent MOSFET threshold voltage extraction methods.* Microelectron Reliab 42:583–596

Chapter 3
Results: NLA Using a Short Pulse Duration KrF Laser

3.1 Introduction

This chapter describes the research performed on samples annealed with the KrF laser from IPG Photonics (USA), which features a short pulse duration of 25 ns, with a wavelength of 248 nm. All the samples analysed in this chapter are based on 50.8 mm wafers and have been implanted in the Faculty of Physics, using the available ion implanter belonging to CAI Técnicas Físicas.

The thesis started with some wafers already implanted and laser annealed that the research group already had in stock. They were used to fabricate and measure the first micrometrical pixels using Ti supersaturated material. We also used two fragments solely for material characterisation. The pixel matrix fabrication route was improved iteratively; some problems were identified and solved, which led to the implementation of new experimental techniques in the research group, as a new multi-cathode sputtering system (Sect. 2.5.4), a modified RIE process (Sect. 2.5.5) and new wire bonding equipment (Sect. 2.5.7).

On the meantime, a new set of samples was designed for the next batch of 50.8 mm wafers, which would conform the bulk of the research contained in this thesis. However, the KrF laser with which we had been working on for the last years was no longer available. Therefore, it was necessary to find another laser source to anneal the already Ti implanted samples. We finally chose a laser with longer pulse duration, which was used for all samples described in the rest of the thesis. All the information of 50.8 mm wafers using the long pulse duration laser is detailed in Chapter 4.

3.2 Material Fabrication

This sub-section describes the steps followed to fabricate the Ti supersaturated Si samples, which are later used to fabricate devices.

D. Montero Álvarez, *Near Infrared Detectors Based on Silicon Supersaturated with Transition Metals*, Springer Theses,
https://doi.org/10.1007/978-3-030-63826-9_3

Prior to Ti implantation, we performed a boron implantation process on the backside of the wafer. The purpose of this ion implantation is to reduce the barrier between the metallic contact and the semiconductor, aiming to achieve an ohmic contact, reducing the contact resistance for pixel transversal measurements. This is usually referred to as Back Surface Field (BSF) in the literature [1]. Boron implanted atoms were activated by means of a RTA process.

All samples in this lot have the same Ti implanted dose, in a two-step ion implantation process: a first one at low energy and 20% of the total dose and a second one at high energy and 80% of total dose. With this double process, we aim to obtain a thicker layer with a rather constant profile. Full details on sample preparation are detailed below, in order of realisation:

Back surface

- Starting wafer material: p-type Si wafer, 1–10 Ω·cm, reference P1.
- Ion implantation process, dose and energy: ^{11}B, 1x10^{15} cm^{-2} and 35 keV.
- RTA process, 900°C, 20 seconds in Argon ambient.

Front surface

- First ion implantation process, dose and energy: ^{48}Ti, 1x10^{15} cm^{-2} and 35 keV.
- Second ion implantation process, dose and energy: ^{48}Ti, 4x10^{15} cm^{-2} and 150 keV.
- Laser thermal annealing process: 1.8 J/cm^2 in 1 mm^2 spots, 10 μm overlapping.

All ion implantation processes on this set of wafers were performed with a tilt angle of 7°. The RTA process was performed before the Ti implantation processes to avoid possible changes in the optoelectronic properties of the Ti Implanted Layer (TIL). Previous studies in our research group suggested that thermal treatments after the laser annealing process could lead to a change on the material properties with Ti atoms [5], and other authors in Te atoms [5] or in chalcogens [5] since supersaturated silicon is a thermodynamically metastable material.

3.3 Material Characterisation

In order to design a functional photodetector based on Ti supersaturated Si, it is important to first characterise and understand the properties of the material. Several published works in our research group have covered most of the material characterisation of Ti supersaturated layers [5–24], so it will not be included here, but in the Annex A for clarification. The reading of Annex A is encouraged for those readers that are not familiar with the compositional, structural and electro-optical properties of Ti supersaturated Si layers. Note that most of the research published in our research group was based on Ti implanted layers on n-type Si substrates. We will consider that the structural and compositional properties of the Ti implanted layers of p-type substrates (used in most of this chapter) are similar to those found on n-type substrates.

In this section, we briefly describe further advances in the comprehension of the material properties performed within the frame of this thesis.

3.3.1 Ruling Out the Origin of the Sub-Bandgap Photoresponse

The origin of the sub-bandgap photoresponse of both non-implanted and Ti implanted samples has been in the focus of our research. We measured that non Ti implanted samples exhibited sub-bandgap response, up to around 0.65 eV, at room temperature. As pointed out in other references [25], this origin could be located in the dangling bond of atoms close to the surface, what has been called surface defects. Provided that the surface defects play an important role in the sub-bandgap photoresponse of non-implanted samples, we wanted to experimentally verify this hypothesis, not only on the reference sample, but also on Ti implanted samples, to see if the sub-bandgap response of the latter could be attributed to surface defects. It is well-known that passivating the surface using silicon oxide may drastically reduce the effect of surface defects in the conduction mechanisms of Si-based devices [26]. Therefore, we measured, using the van der Pauw configuration, the sub-bandgap spectral photovoltage of different samples, before and after depositing a silicon dioxide layer on both the front and the backside of the sample. The properties of the samples used for this study are detailed below:

Substrate: the same used in some of our previous works [23]. Reference N2 substrate in Table 2.1.

Pre-Amorphisation Implantation (PAI): self-implantation of [28] Si atoms at a dose of $5\text{x}10^{15}$ cm^{-2} and an energy of 170 keV.

Double Ti ion implantation: 10^{15} cm^{-2} at 35 keV, followed by another Ti implantation at $4\text{x}10^{15}$ cm^{-2} and 150 keV.

NLA process using the KrF laser source at 1.0 and 1.4 J/cm^2, 1 mm^2 with 10 μm overlapping.

The samples used here are similar to the ones used by Garcia-Hemme et al. [23], experiment described in Fig. 8.4 on Annex A. However, before the Ti implantation process, we performed a PAI of Si atoms at higher energies, aiming to produce an amorphous layer thick enough to minimise channelling for the Ti implantation process. The aim of this PAI process is to produce steeper Ti profiles. In his thesis, García-Hemme [27], linked the magnitude of the sub-bandgap photoresponse of Ti implanted samples to the presence of defects coming from the so-called "implantation tails". The implantation tails are caused by Ti atoms that are not in concentrations high enough to reach the IB formation limit. Therefore, instead of forming an impurity band, Ti atoms may act as recombination centres, possibly blocking the generated carriers inside the IB from reaching the substrate and participating in the conduction mechanisms. Therefore, for this study, we chose PAI samples to reduce the thickness of the implantation tails, aiming to increase the EQE, which would potentially allow the detection of sub-bandgap photons at room temperature.

Square-shaped samples of 1 cm^2 were cut, followed by the deposition of four triangular metallic contacts in the corners, prior to the measurement process. The electrodes are composed of 50 nm of Ti followed by 100 nm of Al, being aluminium the metal in contact with air and Ti with the substrate. The samples were placed in the sample holder displayed in Fig. 2.6, using silver paint to connect the copper wires to the former. We injected a current of 1 mA and we set the chopping frequency to 87 Hz for all photoconductance measurements.

After measuring the samples, with and without Ti implantation, we remove the silver paint using acetone, which are then cleaned using IPA. The sample is submerged in a 1:50 mixture of HF:H_2O during one minute to remove the native oxide, inside the transfer chamber of the silicon dioxide deposition cluster, the previously described ECR-CVD (Sect. 2.5.3). The chamber was previously filled with nitrogen to avoid the re-oxidation of the sample. To avoid the wet etching process of the metallic contacts, we deposited drops of photoresist over the contacts, as photoresist is barely soluble on HF. The sample is then placed inside the vacuum chamber. The silicon dioxide deposition process was performed at a fixed temperature of 100°C during 40 minutes. The flux of oxygen and silane were set to 5 sccm and 8 sccm, using a RF power of 100 W. The total deposited thickness was 60 nm. After the first deposition process on the front surface (the one with the Ti implantation and the contacts), we flipped the samples in nitrogen atmosphere and repeated the same deposition processes on the back surface. After the deposition, the samples were submerged in DMSO to eliminate all the photoresist, which were later measured in the IR photoresponse cluster.

Examining the non-deposited samples, both the non-implanted and the Ti implanted samples exhibit sub-bandgap response. In the case of the non-implanted sample, there is a peak at around 0.62 eV (2000 nm), which we have identified as a second-order of the diffraction grating in our experimental set up. The peak is due to parasitic light coming from the diffraction process, whose real energy is 1.24 eV (1000 nm). The peak may appear only in samples with low sub-bandgap response, but high response at 1000 nm, close to the Si bandgap. At energies lower than 0.62 eV the responsivity of the non-implanted sample decays abruptly, reaching the noise level. The Ti implanted sample, however, exhibits photoresponse up to around 0.5 eV, with a slight better responsivity than the reference at energies lower than 0.95 eV. The same peak at 0.62 eV is also visible in this sample.

However, after the deposition process there is a drastic change on the responsivity levels of both samples. The sub-bandgap photoresponse of the non-implanted sample decreased by more than one order of magnitude. The opposite happens to the Ti implanted sample, where its responsivity increased almost two orders of magnitude after the deposition process. It seems that the passivation produced a beneficial effect of the sub-bandgap generated carriers. The responsivity at 0.8 eV (1.55 μm), a key wavelength in optical fibre telecommunications, increased for the Ti implanted sample, from 18 mV/W up to 590 mV/W after the deposition process, 31 times higher. In fact, the responsivity value at this wavelength, at room temperature, is higher than our previously reported best value, by García-Hemme et al. [23], which was set to 34 mV/W.

While both Ti implanted and non-implanted samples showed similar sub-bandgap responsivity levels up to 0.62 eV before the deposition process, the Ti implanted sample exhibits, after the deposition process, almost four orders of magnitude higher responsivity than the non-implanted sample in the sub-bandgap region.

We measured another Ti implanted sample, using the same implantation parameters, but with lower laser fluence of 1.0 J/cm^2.

In this case, we could not measure the sub-bandgap responsivity at room temperature due to the high noise level present in this Ti implanted sample. We suspect this higher noise level might be due to a higher density of defects after recrystallization [27], since the laser fluence in this sample (1.0 J/cm^2) is lower than the one used in the previous sample, annealed at 1.4 J/cm^2. Therefore, we used the cryostat to lower the temperature up to 250 K, where we measured sub-bandgap photoresponse. The trend is the same as we previously observed in the sample Ti implanted but laser annealed at 1.4 J/cm^2 displayed in Fig. 3.1: the deposition process of a passivating layer enhances the sub-bandgap responsivity.

An improvement of the contact quality after the thermal process during the deposition of SiO_2 is discarded as a possible explanation of the change in the sub-bandgap photoconductance, as it could not explain why the non-implanted sample worsened its photoresponse while the Ti implanted sample increased it. We also considered the increase on incoming photons to the sample due to the Anti-Reflecting Coating (ARC) effect of the deposited layer, which could reduce the reflectance of the sample.

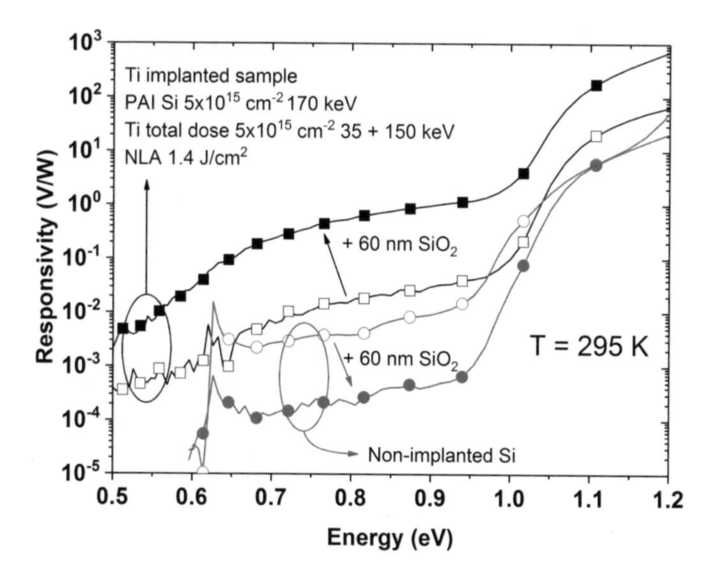

Fig. 3.1 Responsivity of a non-implanted, non-laser-annealed Si substrate and a Ti implanted one, before and after the deposition of an oxide passivating layer at room temperature. Open symbols stand for the bare samples, while solid symbols represent the deposited ones

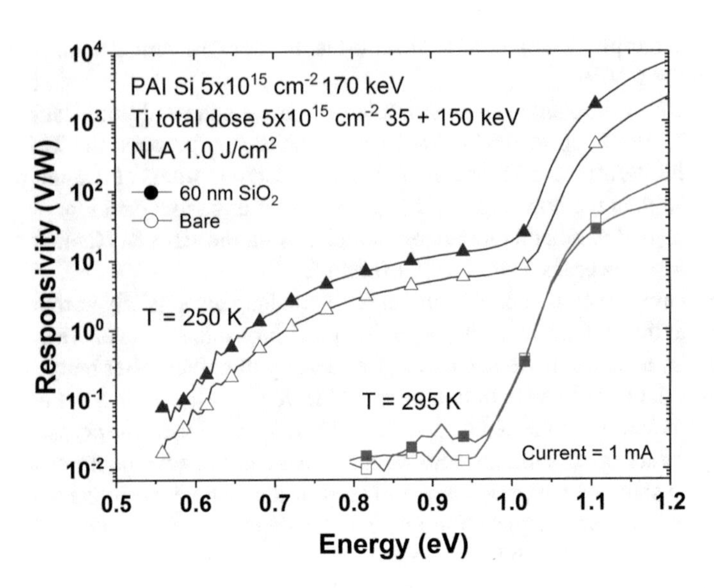

Fig. 3.2 Responsivity as a function of the photon energy, for a PAI processed sample, double ion implanted with Ti atoms at a total dose of $5\text{x}10^{15}$ cm^{-2}, subsequently annealed with a laser fluence of 1.0 J/cm^2. The samples are measured before (open symbols) and after the deposition of an oxide layer 60 nm thick (solid symbols), at two different temperatures

Again, this possibility is discarded as it should have increased the photoresponse of both samples at the same time. The third possibility would involve bulk defects, coming from the ion implantation, but they were already discarded in the study of Olea et al. [21] on self-implanted, subsequently laser annealed Si substrates (Fig. 3.2).

Finally, we propose the possibility of the passivation effect of the deposited layer in the samples. Surface defects are a source of carrier generation, but also of high recombination. Assuming that the sub-bandgap response of the non-implanted samples is coming from the surface defects [25], the passivating layer could reduce the density of defects, thus decreasing the carrier generation that contributes to the photovoltage in the non-implanted sample. The recombination would be lowered as well, but there are no other mechanisms that could absorb sub-bandgap photons, so the net conductivity of the non-implanted sample decreased after the deposition of the oxide layer. In Ti implanted samples, two main sub-bandgap photogeneration mechanisms could coexist: the surface defects and the generation associated to the Ti atoms inside the silicon lattice. The implanted layer is located close to the surface, so the carriers generated by Ti atoms could be recombined close to the surface. After the deposition of the passivating layer, the surface generation process would be minimised, as it happened to the non-implanted sample, but also the surface recombination. The carriers generated by the Ti atoms would increase their lifetime, increasing finally the conductivity of the layer under illumination.

The use of passivation layers is further justified after this experiment, especially in microscale devices, where the surface-to-volume ratio could be higher than in macroscale devices. This experiment also opens the door for improvement of the responsivity, as the chosen thickness is not optimised for the IR range. We propose the use of silicon oxide layers in the order of 250 nm to minimise reflectance in the range of 1.55 μm for future experiments.

Therefore, from the experience acquired in our research group, we started the development of my thesis with an already suitable material for IR sensing at room temperature: its Ti concentration is high enough to supersaturate the implanted layer [19], its crystal quality is good [10], while it is IR absorbent [12]. Besides, the IR absorption process leads to a measurable electrical signal under illumination [23]. All conditions listed above are desirable for the light-absorbing material of a photodetector.

However, as explained at the beginning of this subsection, the work done in this thesis is almost based on p-type silicon substrates, as opposed to the n-type substrates used our previous studies, referenced in this sub-section. The migration from n-type to p-type substrate is justified by previous experiments performed by our two research members E. Garcia-Hemme and J. Olea, which showed photoresponse on photodetectors based on Ti supersaturated layers at room temperature on p-type Si substrates.

3.4 Material Inhomogeneities

Since the beginning of our research on transition metals supersaturated silicon (with Ti, V, Cr or Zr) we have dealt with three different inhomogeneity problems. However, as the sample size was always orders of magnitude bigger than the average inhomogeneity feature size, we expected a negligible effect of the inhomogeneities on our measurements. The same assumption may not be valid when going down to microscale devices, in which the size of the inhomogeneities are in the order of the pixel size.

3.4.1 NLA Overlapping Lines

The first one, especially present on samples annealed with the laser from IPG Photonics, is the effect of the overlapping between adjacent laser shots. In order to quantify the possible effect of these lines in the topography, we use different techniques. We first measured the height profile using the optical profiler described in Sect. 2.3.6.

Figure 3.3 shows the height profile of part of a manufactured device exhibiting several overlapping lines. Line number 1 (indicated by L1 in the figure) is around 570 nm, while line 2 (L2) is only 210 nm high. A high variability has been found between lines of the same sample, and also between samples. This behaviour led us to think that this height, measured using optical interferometry, may be caused

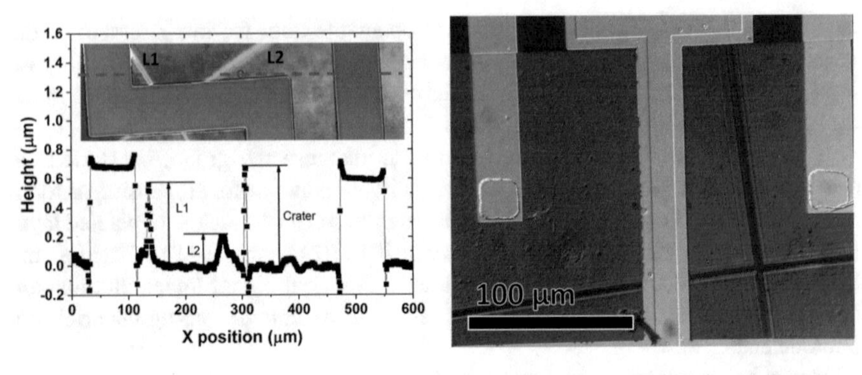

Fig. 3.3 Left: optical profile of two NLA overlapping lines between metallisation tracks. L1 and L2 stand for lines 1 and 2. The inset shows the 2D contour height map, with a dashed red line indicating the cut plane. Right: NLA overlapping lines as seen from an optical microscope

by an optical effect of unknown origin. The use of an optical microscope could not help determining if the height of the lines was real. There may seem to be also diffraction lines, parallel to the NLA overlapping lines, that are not visible using other techniques. To test this hypothesis, we tried other experimental techniques, mainly AFM and cross-section FIB-SEM. Using AFM we could not extract any conclusive measurement: the lines could not be detected. We studied the topography using FIB-SEM technique and tilting the sample 53° to observe the cross-section.

Results from cross-section FIB-SEM may corroborate the negative results coming from AFM: NLA overlapping lines may not produce any significant surface modification. The optical profilometry measurements led to incorrect heights of the overlapping lines. We performed cross-section cuts of several NLA overlapping lines, in different samples with different Ti doses and NLA processes and we could not find any measurable feature corresponding to the overlapping of laser shots. Figure 3.4 shows the worst-case scenario, in which not two, but four different shots were fired on the same area. There might be a depression in the four laser-exposed area, if any, of less than 10 nm with respect to the surroundings, but the exact value cannot be obtained (as displayed in Fig. 3.4). The diagonal of the four times overlapped area is around 30 µm. Due to the reduced field of view of the microscope it was not possible to measure such small difference with enough precision. In any case, it is small enough to possibly not affect the lithography and other fabrication processes related to device fabrication.

With respect to the possible effect from the electrical point of view, we performed several measurements using the point station available at our laboratories (Fig. 2.4). To do so, we placed the point of the needles at a fixed distance between them and proceeded to move the sample perpendicular and parallel with respect to an overlapping line, using several samples with different Ti implanted doses and NLA processes. The results were negative. We could not find any correlation between the distance of the needles, nor the relative position of them with respect to the overlapping line.

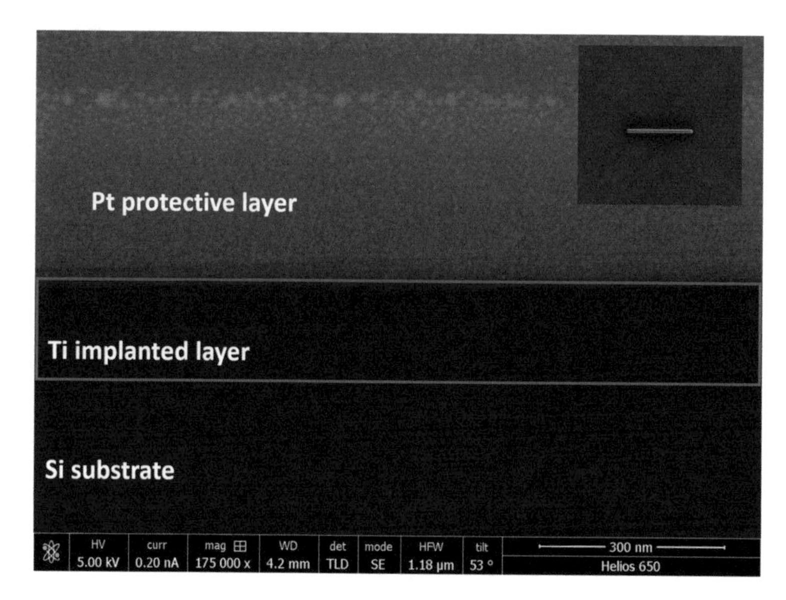

Fig. 3.4 NLA overlapping line cross-section FIB-SEM micrograph. The inset shows the area were the cross-section was obtained. The place is a corner of the laser exposed area, where four different shots overlap

The possible electrical effect of the overlapping lines on the pixel properties is yet to be determined with more accurate techniques.

3.4.2 *Craters*

The second effect are what we have labelled as craters. They are random structures, with the appearance of a crater, possibly produced by an explosive phenomenon during or after the NLA process. We qualitatively studied the possible influence of the Ti dose and the NLA process in the number and size of the craters, using an optical microscope on previously fabricated samples.

Figure 3.5 is composed of several pictures obtained from an optical microscope. The upper row shows the non-implanted samples. The lower row, the samples implanted, using the already described double ion implantation process of Ti atoms at 35 + 150 keV and a total dose of 5×10^{15} cm^{-2}. Each column shares the same laser process, being the first the as-implanted samples, the second the samples annealed at 1.2 J/cm^2 and the third at 1.8 J/cm^2. The as-implanted sample do not display any sign of laser overlapping lines nor craters. At 1.2 J/cm^2 there is not any visible change in the non-implanted sample. Overlapping lines and few isolated small craters are visible for the Ti implanted sample. At the highest fluence the craters are visible in both the Ti implanted and the non-implanted sample. The number of craters is

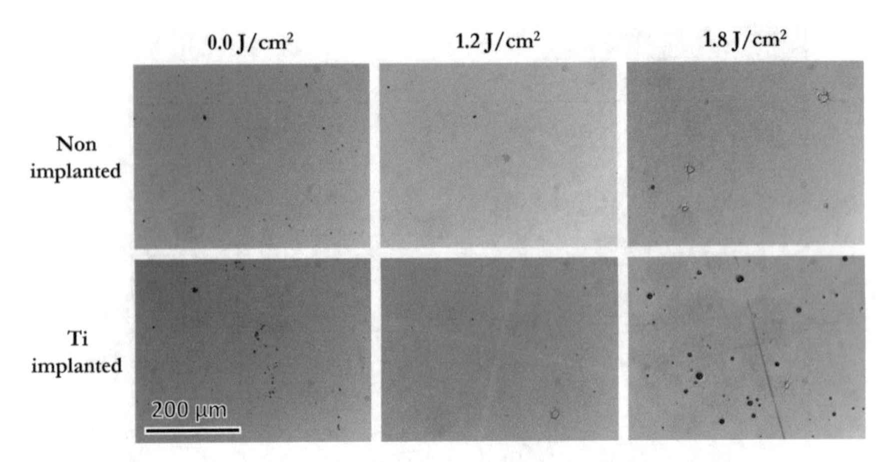

Fig. 3.5 Surface pictures taken using an optical microscope to address the size and quantity of craters. Upper row shows non-implanted samples, while lower row contains pictures of double Ti implanted samples with a total dose of $5\text{x}10^{15}$ cm^{-2}. The scale is shared in all pictures. Some dark points are produced by deposited particles on the surface

increased, along with the average size of the craters. NLA overlapping lines are only visible on the Ti implanted sample. The conclusions extracted from Fig. 3.5 are summarised next:

- Overlapping lines are only visible on Ti implanted samples. This may point to a Ti-related origin of the lines.
- Size and number of craters increases along with the laser fluence for Ti implanted and non-implanted samples.
- The amount of craters is greater on Ti implanted samples with respect to non-implanted ones.

In Fig. 3.3 we could observe a crater using the optical profiler, where we could measure a height of around 650 nm. Other craters were observed using the same technique, giving different values between 50 to 700 nm high, featuring an average diameter of 10 to 50 μm. We used the cross-section FIB-SEM technique to study the morphology of several craters found on several samples.

Figure 3.6 shows the morphology of two craters in different samples, which are representative for most craters found in our study. We have not found any correlation between the laser energy and the crater morphology in the analysed samples. From the upper right Fig. 3.6 we check that the craters effectively deform the surface. The horizontal yellow line indicates the ground level of the surroundings of the crater. We find a peak of around 639 nm on the border of the crater, while there is a depression of around 371 nm halfway towards the centre of the crater. The lower crater shows the typical "wrinkles" we have found in many craters, featuring a wave-like cross-section structure of around 90-130 nm high. In order to see the chemical composition of the craters, we performed EDX measurements with the built-in Ametek or INCA modules in the Helios 600 and 650, respectively, of the FIB-SEM instruments.

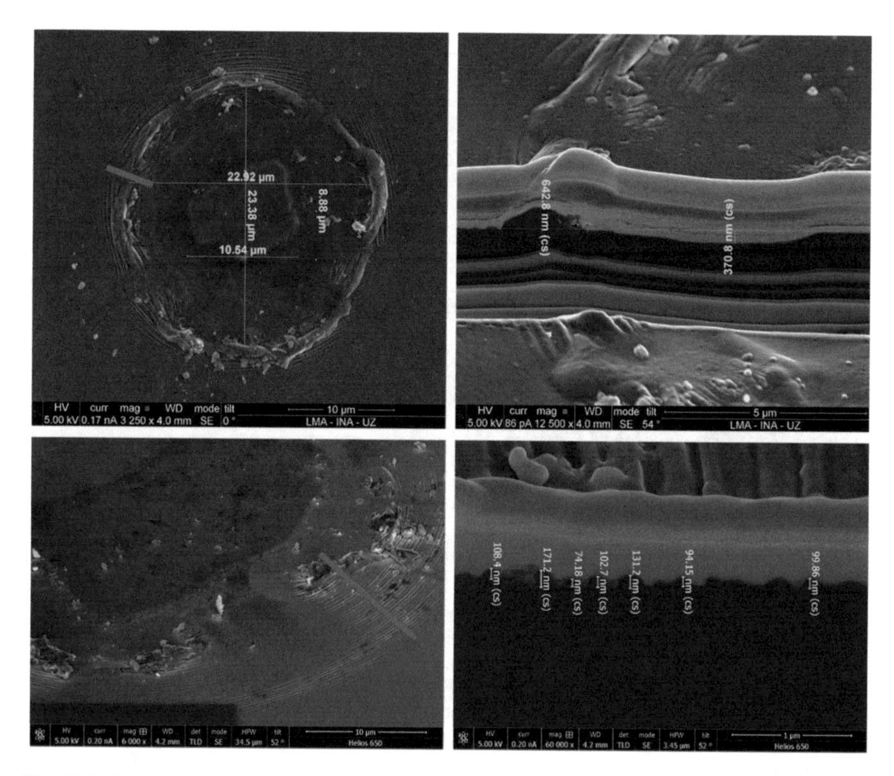

Fig. 3.6 Cross-sectional FIB-SEM micrographs of craters found in double implanted Ti samples laser annealed at 1.0 J/cm^2 (upper row) and 1.4 J/cm^2 (lower row). Red lines indicate the cut plane

Figure 3.7 displays the chemical composition as obtained by EDX in several points of the crater. Point A is placed in the middle of the crater, whose EDX shows mostly silicon. Point B is located outside the crater; the result is almost the same as inside the crater. Points C and D are placed in what may look like foreign particles, in the size of a few microns, that are somehow attached to the wafer, which have always been found in the surroundings of the craters. Their chemical composition is different, finding always oxygen apart from silicon, and, in some cases, further elements as in point D, containing carbon and also metals as Ag, Ca or K. These measurements seem to point out to the possible nature of the crater formation process, which could be the influence of previously deposited particles in the surface before the NLA process. We may discard the deposition of the aforementioned particles after the NLA process, as we cleaned the samples with organic solvents and ultrasounds before performing FIB-SEM and EDX analysis.

Ti atoms have not been detected in any measurement. The Ti concentration in the surface, as seen by SIMS on Fig. 8.1 (Annex A), may reach up to 2–3x10^{20} cm^{-3}, which is around 1% of concentration, in the limit of the EDX measurement. In the craters shown in Fig. 3.6 we extracted an irregular surface profile, with peaks and valleys, of several hundreds of nanometres. Taking into account the available data

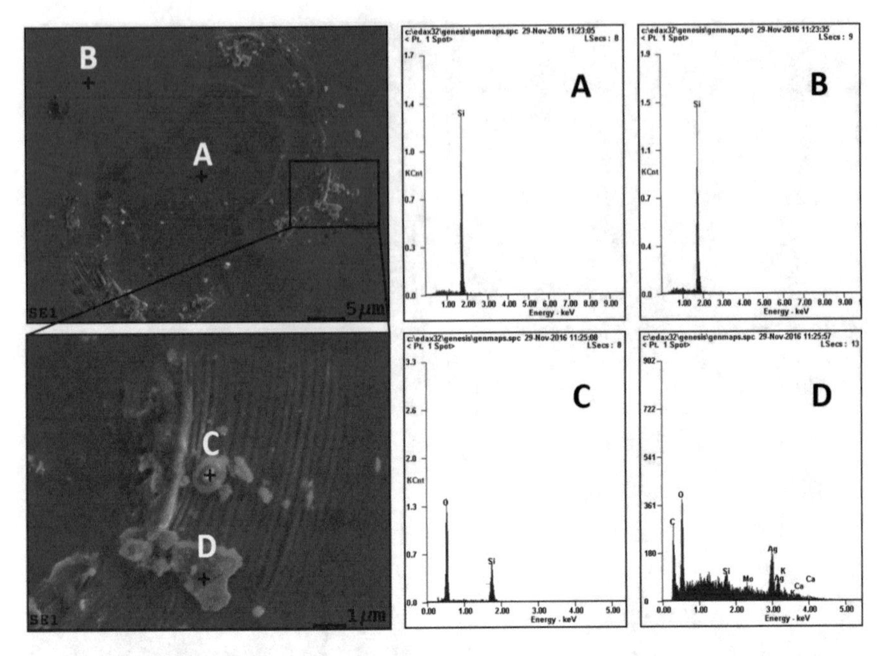

Fig. 3.7 EDX measurements on different points of the same crater, of a sample double Ti implanted subsequently followed by a NLA process of 1.2 J/cm^2

from SIMS, the thickness of the Ti supersaturated layer could be in the range of 100–200 nm. Therefore, we cannot conclude if craters are affecting the continuity of the Ti implanted layer.

Judging from available data, we propose for the next set of samples to perform an exhaustive surface cleaning process to minimise the density of deposited particles prior to NLA process.

3.4.3 Cellular Breakdown Structures

Cellular Breakdown (CBD) structures can be considered an inhomogeneity as well, although its feature scale is considerably lower than that of the overlapping lines or the craters. Some previous discussion on CBD structures can be found in Annex A. Here, we examine the CBD present in the Ti supersaturated material in an already fabricated mesa device, with an Al deposited layer atop Si. Instead of using XTEM to obtain the micrographs, as in Fig. 8.2, we use SEM in cross-section, done by ion milling using the Helios FIB-SEM instruments (Fig. 3.8).

Using EDX we could estimate the chemical composition of the CBD structures. We could not find any trace of Ti atoms, nor in the columnar structures, nor in the surroundings. The main signals coming from these regions is silicon, along with small traces of aluminium and platinum. The surface does not seem flat; the deposited aluminium seems to fill the small trenches found in the termination of the columnar

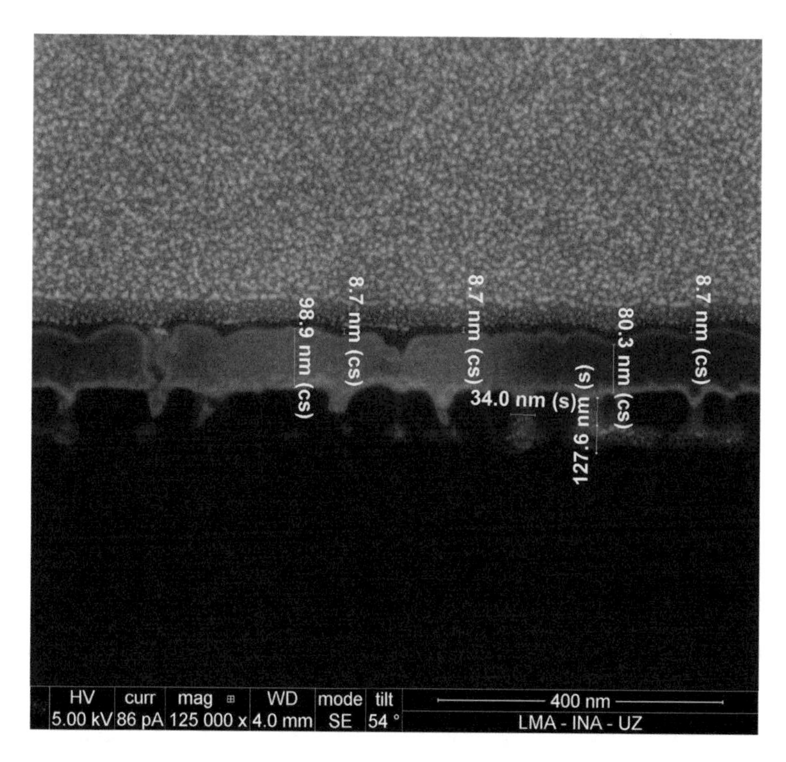

Fig. 3.8 Cellular Breakdown structures under an Al metallisation layer of a double Ti implanted sample with a total dose of 5x10^{15} cm^{-2} and a NLA fluence of 1.8 J/cm^2

structures. The average size of the CBD structures in the examined samples is around 70 nm deep, while the distance between the columnar structures is in the order of 50–80 nm. The depth value is different from the one obtained by Olea et al. [12], from XTEM micrographs represented in Fig. 8.2 of Annex A, whom obtained around 120 nm. The difference may be in the technique used to obtain the micrograph. During this thesis, we did not measure XTEM samples of the double Ti implanted samples at a total dose of 5x10^{15} cm^{-2}. Therefore, we rely on the previously cited work of Olea et al. to address the crystal properties of the material.

Cellular breakdown structures are also visible from the surface, without the use of cross-section techniques, as ion milling or through TEM lamellas. Using SEM it is possible to observe the cellular structures from above (Fig. 3.9):

The inhomogeneity is observable by mosaic-like structured surface, with an average feature size of 85 $\pm$ 15 nm. The different contrast observed between different cells or grains cannot be experimentally attributed to different Ti concentrations; EDX measurements did not show any Ti content on the surface, as the Ti concentration may be lower than the sensitivity of the EDX technique. Unfortunately, SIMS, whose sensitivity is higher, is neither suitable, as the ion counting is performed in a surface considerably higher than that of the CBD structures.

Summing up, the inhomogeneities found on Ti supersaturated layers may play a role in microscale device fabrication, although the extension of this influence may

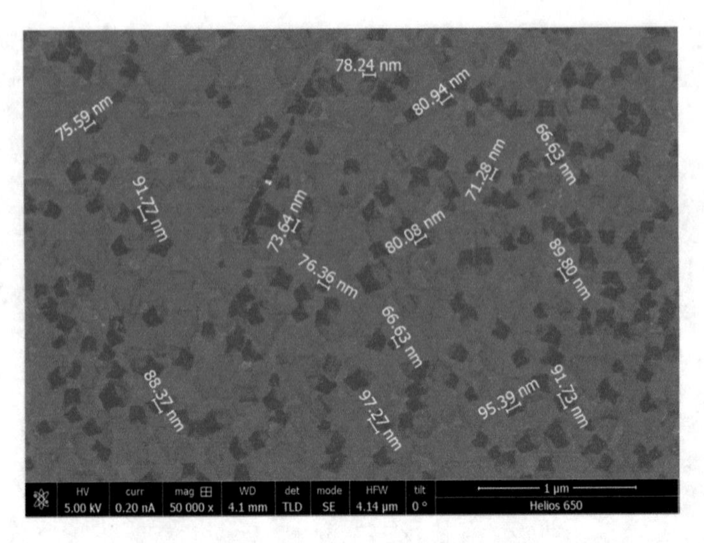

Fig. 3.9 Cellular Breakdown pattern observed by SEM. The sample is double Ti implanted at a total dose of $5\text{x}10^{15}$ cm^{-2}, laser annealed at 1.8 J/cm^2

require further examination. Overlapping lines do not exhibit any pronounced profile feature, as the optical profilometry measurements induced us to think in the first place, while their electrical influence could not be measured. The effect of the craters may play a more important role in the device functioning, as their size is in the order of the device size. From EDX measurements we could not find out if the craters pierced the Ti implanted layer, which leaves the door open for a possible short-circuiting of the implanted layer with the substrate in case of metal deposition over the craters. It seems that crater formation process may be driven by solid particles present in the surface at the moment of the NLA process, judging from EDX measurements. Finally, the cellular breakdown structures are present in our samples of this subsection. We do not expect to find any significant influence of these structures in the homogeneity of the devices, as their feature size, in the nanometre scale, is orders of magnitude smaller than the pixel size, in the range of several hundreds of microns.

3.5 Pixel Matrix Device Fabrication

The idea of fabricating a pixel matrix, in the micrometre scale, comes from our willing to transfer the Ti supersaturated Si material to the industry, in the form of a commercial Focal Plane Array CMOS Image Sensor to obtain images in the NIR, due to the amount of possible applications of this kind of photodetectors in the industry (see Chapter 1). To reach a maturity level in the Ti supersaturated material high enough to be transferred to the industry, there are several intermediate goals that must be achieved. First, we want to test the suitability of the material in the microscale. So far, our previous experiments have been based on relatively big devices, in the millimetre or centimetre scale. There could be inhomogeneity issues that could slip by on macroscale devices, but could impose a problem in smaller devices. Secondly, small, more complex devices could enhance the characterisation process of the material, like noise level or responsivity itself. For example, in the case of low mean-free path carriers, the distance between the generation site and the contact (collection site) may be determinant in the performance of the device. In this scenario, smaller devices could be advantageous over bigger alternatives. The best example of this cunning is the research done on the metallisation configuration in solar cells [28], where the electrode configuration affects the final efficiency of the solar cell. Another reason to go for the microscale devices could be to avoid the alternative conduction paths or leakages that may occur in the borders of the sample. Depending on the device structure, the effect of the borders in the measurement could be minimised, for example, by doing mesa structures or using guard rings. Finally, it may be interesting to test the material compatibility in different fabrication steps as, for example, in dry, wet etching processes or deposition processes. The feedback coming from these experiences may be crucial to develop a device fabrication route compatible with the supersaturated material at a commercial level.

However, not all the features are advantageous. The inherent complexity of the fabrication process must be taken into account: depending on the complexity of the device, its fabrication route could be composed of, at least, tens of different steps, which need to be individually controlled. The chance of making a mistake is increased accordingly. Besides, there is another problem, inherent to the microscale, which is the cross-talk, i.e., the signal received by a pixel when one of its neighbour is being illuminated. This signal degrades the performance of the device, and it is usually considered as noise. Cross-talk is inexistent in devices that occupy the whole semiconductor area.

Summing up, it is more time consuming and more expensive to manufacture microscale devices, but it is a necessary step in order to transfer our current knowledge to the industry. Besides, it is possible to obtain new parameters of the material which could be difficult to achieve with macroscale devices.

3.5.1 Device Structure and Layout

The development process of the device structure is complex, and it usually starts by posing several questions about the purpose of the device: a first pixel prototype in the microscale. This decision defines the first designs of the device structure, including dimensions, shape, passivation and metallisation layers. Once the device has been designed, it is possible to start decomposing the final structure into its different steps, which may define the order of the different deposition or etching processes, along with the necessary photolithography masks.

We aim to fabricate a prototype pixel structure, with the following characteristics:

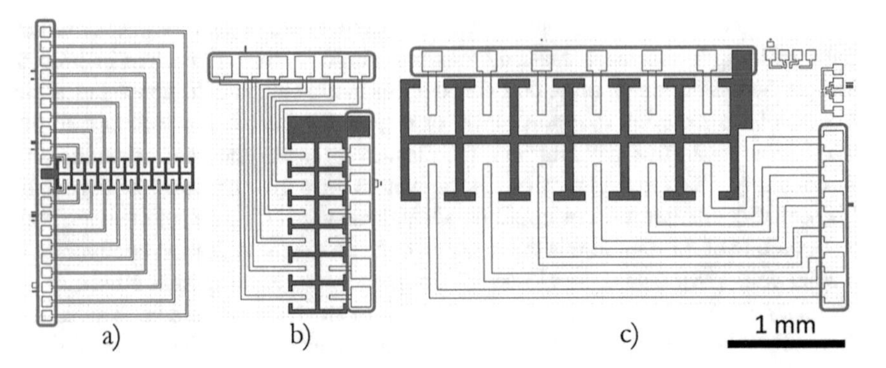

Fig. 3.10 Pixel matrix configuration: (**a**) Small matrix, (**b**) Medium matrix and (**c**) Big matrix. Shaded areas indicate the guard ring contact. The delimited areas by red rectangles indicate the pads used for wire-bonding to the chip carrier

- Time and cost effective. Simple enough to minimise the source of errors, with the possibility to be manufactured inside our laboratories.
- Pixel matrix. Easy integration within a pixel matrix to study the homogeneity of the supersaturated layer. We chose square-shaped devices of several hundreds of microns of lateral size. The pixel matrix is formed by two rows to reduce the number of metallisation layers.
- Different pixel sizes. Measuring 120 (small pixels), 240 (medium pixels) and 480 μm (big pixels) on the side.
- Back contact. It enables the possibility to transversally measure the pixel structure.
- Isolation structures between pixels. Used to reduce cross-talk. We designed guard ring structures and trenches. The guard ring, without the trenches, can be used for horizontal characterisation.

In total, small matrices have 20 small pixels, medium matrices have 12 medium pixels and big matrices have 12 big pixels. Other dimensions of the pixel, e.g. metal width or separation between pixels is scaled to the pixel size. For example, separation between pixels is set to 20 μm on small pixels, but for big pixels it is 80 μm (Fig. 3.10).

3.5.2 Manufacturing Route

The final structure has been defined. The next step is to elaborate the fabrication route that may lead to the designed structure. During the development of my thesis, up to six different generations of devices were designed and fabricated. We call "generation" a group of wafers which were fabricated using the same manufacturing route. During and after the manufacturing and characterisation processes, some problems may be identified. Changes are proposed for the next generation to overcome them: the development process is continuous and cyclic. The first generations were based on non-implanted wafers; they were used for testing the technology and the fabrication process. Once a minimum level of maturity was reached, we started including Ti implanted wafers in the fabrication route. Here, only the last generation of devices is described for the shake of clarity. More details on the manufacturing route, its previous versions and the proposed changes may be found in my Master Dissertation [29].

The fabrication route has to deal with a characteristic of the already Ti implanted wafers, based on the P1 substrate (Table 2.1): they are Single Side Polished, and the Ti implantation was performed on the polished surface (front surface in the following). It forces the back contact to be on the non-polished surface, the backside. Ti implantation in the front side may impose a problem in cross-talk: the Ti supersaturated layer may electrically behave as a metal [14]. Therefore, the fabricated devices could be short-circuited between them, rendering the whole pixel fabrication process useless. The solution we took was to create a discontinuity in the supersaturated layer, by using a dry etching process, which would create what has been called "mesa" devices as the first step. However, as we want to selectively etch the surface, we need to do a photolithography process first. More details on the full photolithography process, from initial cleaning process to development and photoresist elimination can be found in Annex B.

Colours of the legend: light grey represents silicon, dark green the Ti Implanted Layer (TIL), violet the positive photoresist (PR +), black the mask, cyan the silicon nitride (SiN_x), magenta the negative photoresist (PR -), dark grey is aluminium and finally yellow stands for gold. Figure 3.11 shows the layout of the first step, the lithography process to create the mesa structure. The upper row is the top view of the device, while the lower row depicts the cross-section in the middle of the structure, indicated by the dashed red line in Fig. 3.11a. In b), positive photoresist is applied by using the spinning coating method. The photoresist used is an AZ4533 from Microchemicals, and it is expected to provide a uniform thickness of 3.3 μm when

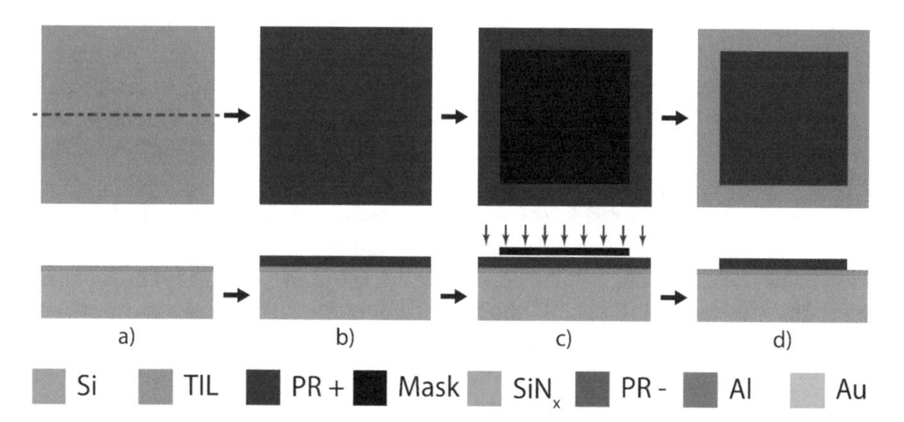

Fig. 3.11 Photolithography process used to create the mesa structure. (**a**) Starting wafer. (**b**) Applying photoresist. (**c**) UV exposure through mask. (**d**) Development of photoresist

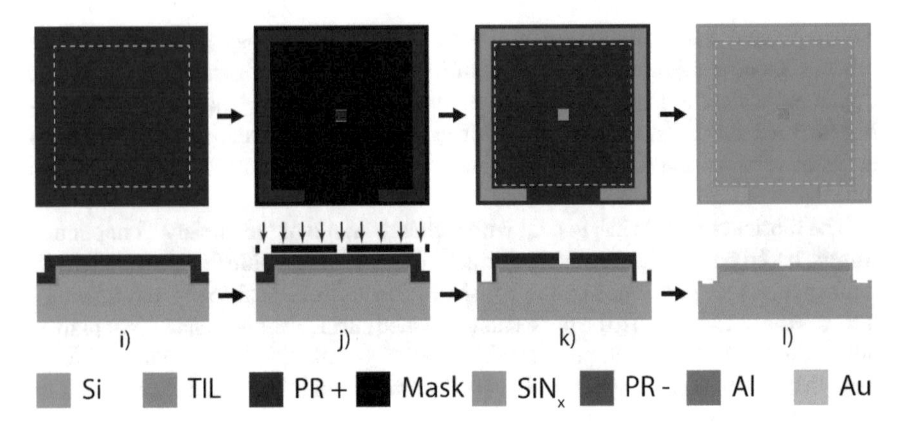

Fig. 3.12 Dry etching and dielectric deposition steps. (**e**) RIE process. (**f**) Mesa structure etched. (**g**) Photoresist wet etching. (**h**) Silicon nitride deposition. White dashed line indicates the location of the mesa structure

spun at 4000 rpm during 30 s. In step c), the photoresist is cured using UV radiation. As it is the first lithography process, it would not be necessary to align the mask, however, to ease the ulterior cutting process of the sample, we chose to align the mask with respect to the flat of the wafer. In step d), we develop the photoresist using the developer AZ 826 MIF from Microchemicals. The motif of the mask is transferred to the photoresist, which is now ready to be dry etched in the RIE equipment (Fig. 3.12).

RIE process is performed in our homemade RIE system, at a fixed RF power of 10 W, at 0.1 mbar of pressure, during 4 minutes. The etching rate at these particular conditions is around 375 nm/min, giving a final mesa height of around 1.5 μm. After the RIE process, the remaining photoresist is removed using an organic solvent, in our case, DMSO. Then, we etch the native oxide inside the transfer chamber of the ECR-CVD, where we later deposit a clean layer of silicon nitride, which will be used for passivation, whose importance has been pointed out in the previous sub-section, and to act as an ARC layer. The deposition process is performed at a RF power of 100 W, with a gas flux of 4 sccm for N_2 and 8 sccm for SiH_4, during 120 minutes at room temperature to deposit a final silicon nitride layer 210 nm thick. We chose room temperature deposition process to reduce the thermal budget of the Ti implanted layer (Fig. 3.13).

The next step is the deposition of another layer of positive photoresist to define the nitride opening areas or "pad openings". The photoresist is cured by UV radiation through the aperture mask. In this step, we carefully align the previously defined mesa structures with the mask. Later, the photoresist is developed. Finally, we perform a wet etching process of the nitride layer using BOE (Buffer Oxide Etchant), as described in Sect. 2.5.6, during 4 minutes. Now, there are two areas of semiconductor exposed: the centre contact on the supersaturated layer and the guard ring, where bare silicon is exposed. The next step is metallisation. From previous experiences in the first generations of Silfrared devices, we found advisable to do a lift-off process in this

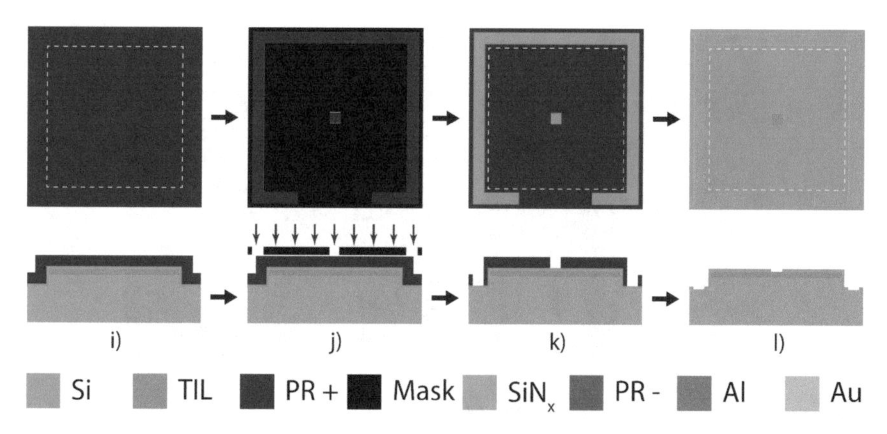

Fig. 3.13 Photolithography of pad opening process. (**i**) Applying photoresist. (**j**) UV exposure of photoresist. (**k**) Development of photoresist. (**l**) Dielectric wet etching

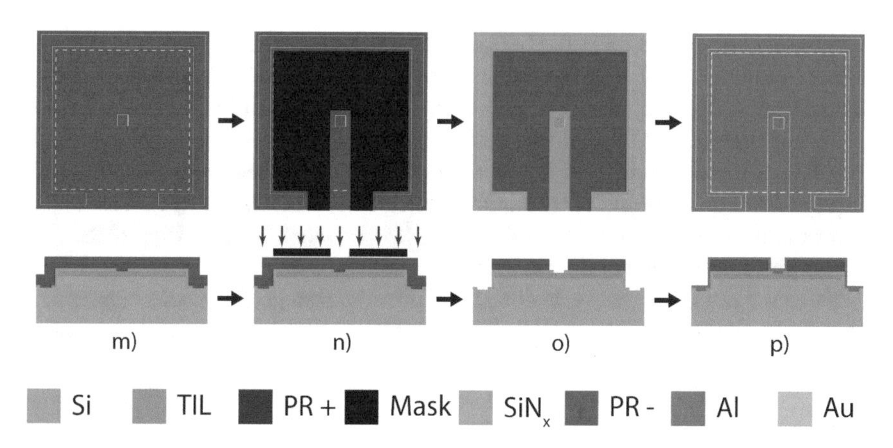

Fig. 3.14 Metallisation photolithography process. (**m**) Applying photoresist. (**n**) UV exposure through metallisation mask. (**o**) Development of photoresist. (**p**) Metal stack deposition

step instead of a wet etching process. Thus, a negative photolithography process goes next in the manufacturing route.

In negative photolithography processes we use the AZ 2035 photoresist from Microchemicals. It must be spun at 3000 rpm during 30 s to provide a uniform thickness of 3.5 μm. After exposure and development, we deposit the metallisation stack, which can be 50 nm of Ti followed by 100 nm of aluminium, or a single Al layer of 150 nm. Both possibilities are deposited by e-beam evaporation, at a constant rate of around 1.2 Å/s (Fig. 3.14).

The lift-off process, using the same organic solvent than for positive photoresist, removes the exceeding aluminium from the sample, revealing the final structure of

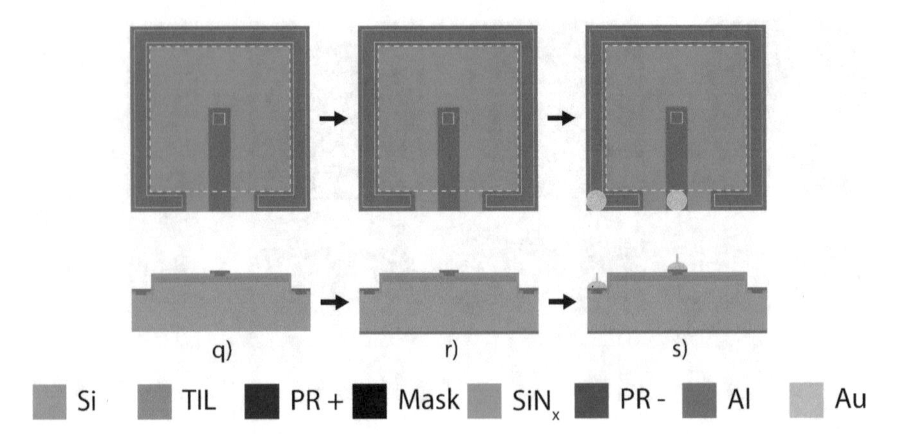

Fig. 3.15 Lift-off and final preparation of the device. (**q**) Lift-off of metallisation layer. (**r**) Backside contact deposition. (**s**) Wire bonding

the device. The next step is the backside metallisation process, which is done by e-beam evaporation of a stack composed of a 50 nm thick Ti layer followed by a 100 nm thick Al layer. The final step of the fabrication route is the wire-bonding process of the sample to the chip carrier. Note that the welding point showed in Fig. 3.15 is merely indicative. The wire bonding is performed on the metallic pads displayed in Fig. 3.10. In the case of the back contact, it is directly contacted using silver paint to the chip carrier surface, which is covered in gold. One of the chip carrier pins is contacted to the substrate.

Once the device layout is designed, we need to arrange the pixel matrices into groups for sample fabrication. For compatibility matters with our characterisation instruments, we decided to design a 1 cm^2 square-shaped die, containing two matrices of each size, completing 88 pixels per die. Each device is univocally identified with a contact number. Included in the same die, we also designed alignment patterns, which would help in the alignment process of successive lithography processes. Several structures to measure the conduction properties of the implanted layers were included using the van der Pauw configuration and the Transfer Line Method (TLM) [30]. Finally, the logo of the project is included, with the acronym Silfrared, formed from the words Silicon and Infrared. To avoid the possible influence of the borders in the sample fabrication, we designed a mask containing four 1 cm^2 dies in the centre of the wafer (Fig. 3.16).

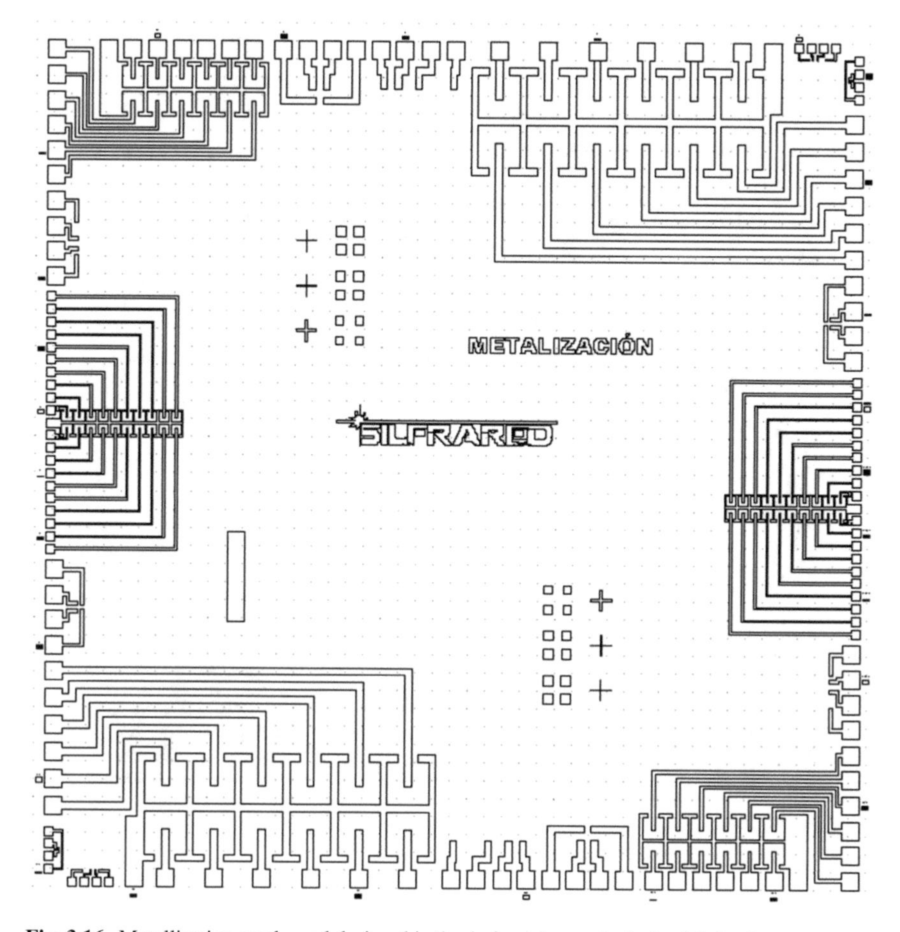

Fig. 3.16 Metallisation mask used during this thesis for microscale device fabrication

After the wire bonding process, the devices are ready for characterisation (Figs. 3.17 and 3.18).

Fig. 3.17 From wafer to device characterisation. The whole process might last up to two weeks

Fig. 3.18 Silfrared die mounted and wire-bonded to the chip carrier. The NLA spots are visible in the purple area (colour caused by the deposited ARC nitride layer)

3.6 Pixel Matrix Process Characterisation

The previous section contained the layout and the fabrication route of the pixel matrix developed in this thesis using UCM laboratories. Here, we summarise part of the characterisation of the fabrication processes.

3.6.1 Optical Profilometry

We use optical profilometry to measure the height of the RIE processes performed for the mesa fabrication step.

From Fig. 3.19 we can extract the mesa height. In this particular case, the RIE was slightly longer than usual, 4 min and 20 s, which led to an etched depth of 1.60 $\pm$ 0.04 μm and an etching rate of 368 nm/min.

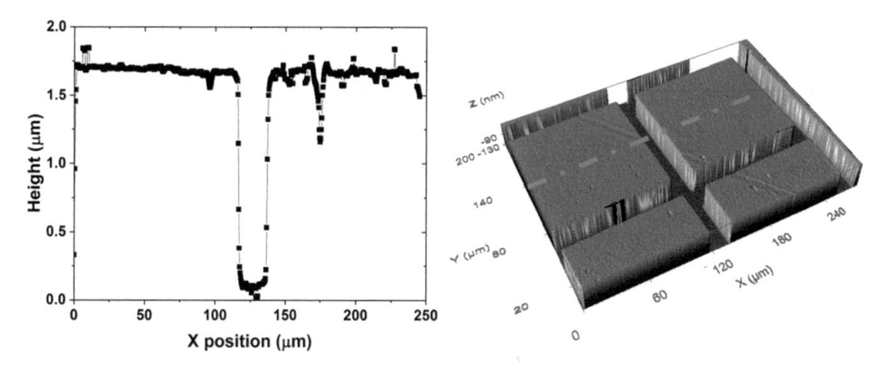

Fig. 3.19 Optical profile obtained using the Wyko NT1100. The RIE process was performed during 4 min and 20 s at 10 W and 0.1 mbar of SF_6 pressure. Right: 3D topography map. Green line indicates the cut plane where the 2D profile was measured, which is shown on the left

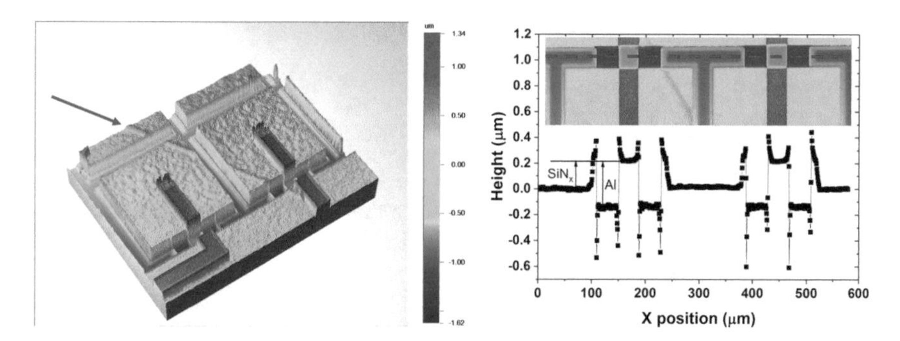

Fig. 3.20 Optical profile of a medium size pixel, obtained using the Wyko NT1100. Left: 3D representation. Right: 2D profile, whose cut plane is indicated in the inset by a red line

It is possible to obtain the thickness of the dielectric and the metallisation layers using optical profilometry. Apart from the strict numerical values of the thicknesses, we also used this technique to visually inspect the quality of the lithography processes.

Left Fig. 3.20 shows the three-dimensional view of two medium-size pixels (240 μm width). The red arrow indicates the NLA overlapping line between two adjacent pulses. In Fig. 3.20 right we observe the cross height profile of the cross-section marked by the dashed red line on the inset. The sharp peaks and valleys close to the edge of each layer are an artifact of the measurement. The overall quality of the photolithography processes is very good, based in the measured dimensions of the pixel (see Table 3.1).

The experimental values are close to the designed pixel dimensions, with relative variations lower than 3% in the worst case, and lie within the expected uncertainty of the combination of deposition, etching and photolithography steps. The thickness of the nitride layer is fully compatible with the experimental uncertainty, but not in the case of the aluminium deposited layer. The targeted value, for this device, was 100 nm

Table 3.1 Features of a medium size pixel, obtained from optical profilometry

Feature	Targeted (μm)	Experimental (μm)
Pixel width	240	241.6 ± 0.5
Pixel separation	40	39.0 ± 0.2
Metal width	40	41.3 ± 0.5
Nitride thickness	0.220	0.221 ± 0.005
Al thickness	0.100	0.361 ± 0.011

of Al, based on the value coming from the e-beam evaporator, but we measured 361 nm instead. This difference, by more than a factor of 3, is not explainable by its uncertainty. We suspect it may be due to the big difference in reflectivity between the nitride and the metal lines. As the technique is based on optical interferometry, there could an induced error when measuring two materials of different reflectivities close between them. Therefore, we may not rely on this technique to measure the thickness of metallic layers atop other materials, as silicon or dielectric layers.

3.6.2 Ellipsometry

Nitride and oxide layers were deposited during this thesis. We use ellipsometry, as described in Section 3.2.3.6, to measure the refractive index and the thickness of the deposited dielectric layers. The refractive index measured corresponds to a fixed wavelength of 632 nm, performed at several points along the deposited wafer.

Figure 3.21 shows the 2D contour map of thickness (left) and refraction index (right) of a SiN_x deposited layer 100 nm thick on a 50.8 mm wafer. We set a total of 49 points homogenously distributed in the wafer, with an edge exclusion of 10 mm to avoid border effects. We use the standard deviation over the average value as the uncertainty of the measurement. Several thicknesses were deposited in this thesis for testing or device fabrication, which are indicated in the Table 3.2.

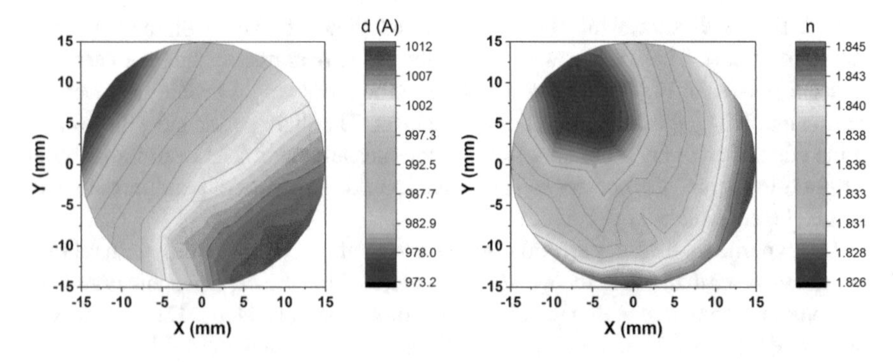

Fig. 3.21 Contour map of thickness (left) and refraction index (right) of a SiN_x layer 100 nm thick

Table 3.2 Thickness and refractive index of dielectric deposited layers used in this thesis, measured using ellipsometry at a fixed wavelength of 632 nm

Dielectric	Thickness (nm)	Refractive index n
SiN_x	100 ± 10	1.84 ± 0.02
	220 ± 20	1.75 ± 0.09
SiO_2	63 ± 8	1.45 ± 0.05
	245 ± 23	1.47 ± 0.06

The refraction index of silicon nitride is dependent on the nitrogen content of the layer, hence its generic SiN_x denomination, instead of addressing the stoichiometry nomenclature. From the studies of Pereira et al. [31], whom deposited nitride layers using the ECR-CVD technique, they obtained lower refractive indices when increasing the nitrogen content. Their lower reported value is 1.88, which is close to our experimental value, indicating that our deposited layers may be nitrogen rich. With respect to silicon oxide values, in the review published by Kitamura et al. [32], the refractive index for silicon oxide is in the order of 1.47 ± 0.02, value that is in accordance to our experimental values.

3.6.3 FIB-SEM

We used FIB-SEM to further study the structure of the device. We etched trenches in several parts of the pixel structure using the ion milling technique, and examined the resulting cross-section cut. The sample was tilted 54° to better examine the cross-section.

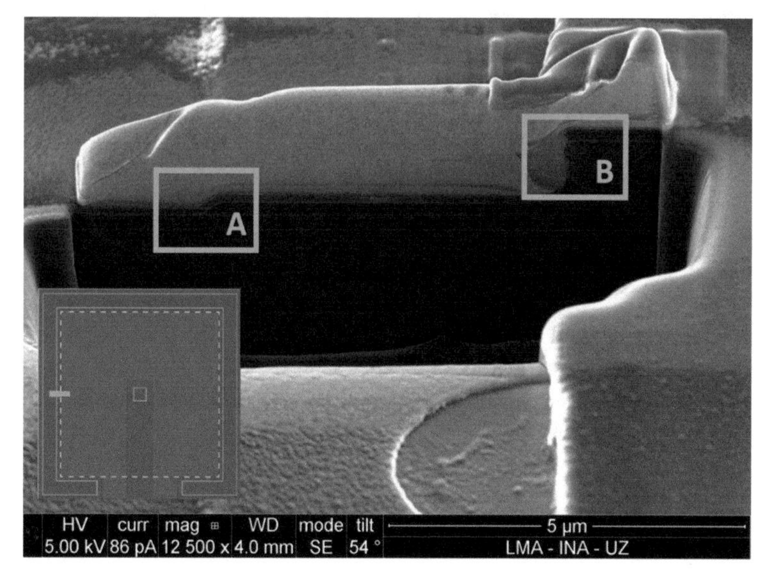

Fig. 3.22 Cross-section of a small pixel, obtained by FIB-SEM, using a tilt angle of 54°. The cut point is shown by a red line in the inset. Two areas of interest, A and B, are indicated

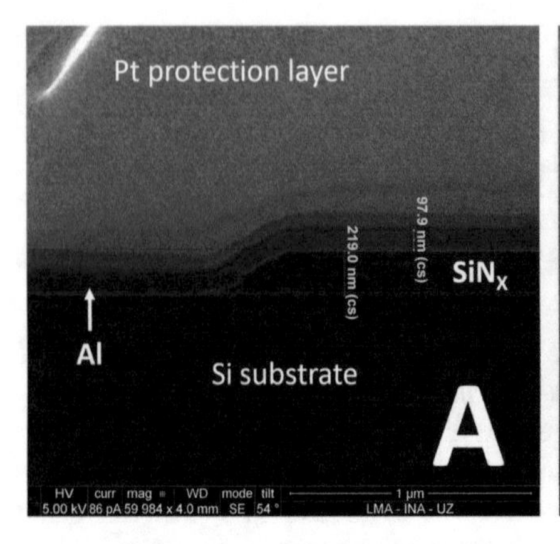

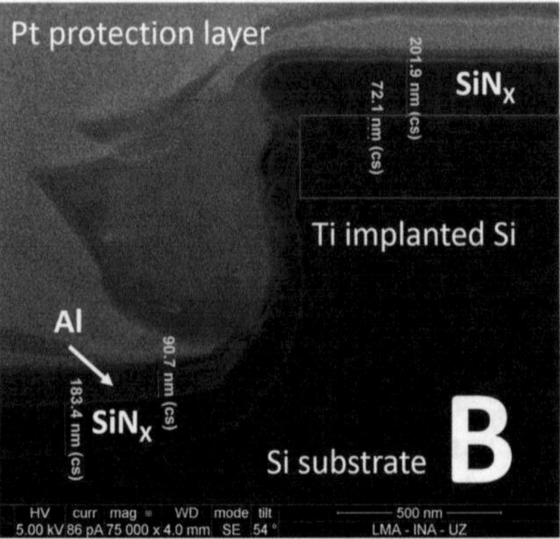

Fig. 3.23 Detailed cross-section on areas A and B indicated in Fig. 3.22. Red box estimates the Ti implanted area

Table 3.3 Average thickness of deposited layers using the FIB-SEM cross-section micrographs

Layer	Targeted (nm)	Experimental (nm)
Nitride	220	219 ± 5
Aluminium	100	98 ± 5

Figure 3.23 A shows the cross-section of the bare Si contact area, located in the trench between pixels. From this picture we could obtain the average thickness of each layer (Table 3.3).

The thicknesses of both layers, experimentally measured by cross-section, are fully compatible with the targeted values, considering the uncertainty. The aluminium layer thickness, measured by means of FIB-SEM, corresponds to the value we were expecting for, as opposed to our previous measurement by means of optical profilometry, which showed a thickness substantially higher. Therefore, FIB-SEM is recommended to provide more accurate results on the thickness of metallic deposited layers.

Figure 3.23 B shows a close view on the mesa structure. The mesa profile is steep, showing that the chosen RIE process may have been highly anisotropic. The steep profile may have caused an interruption in the nitride deposited layer, as it is not conformal, visible by non-communicated nitride grains in the mesa walls. The area enclosed by the red rectangle approximately delimits the Ti implanted area. There are filamentary structures which we interpret as cellular breakdown structures, which extend up to around 72 nm deep.

3.6.4 Optical Microscopy

During the fabrication process we examined the samples regularly to check that the processes were performed according to the desired specifications. There was one particular effect, non-desired, that we identified only on Ti implanted samples, after the etching of the silicon nitride used as passivation and ARC layer (Fig. 3.24).

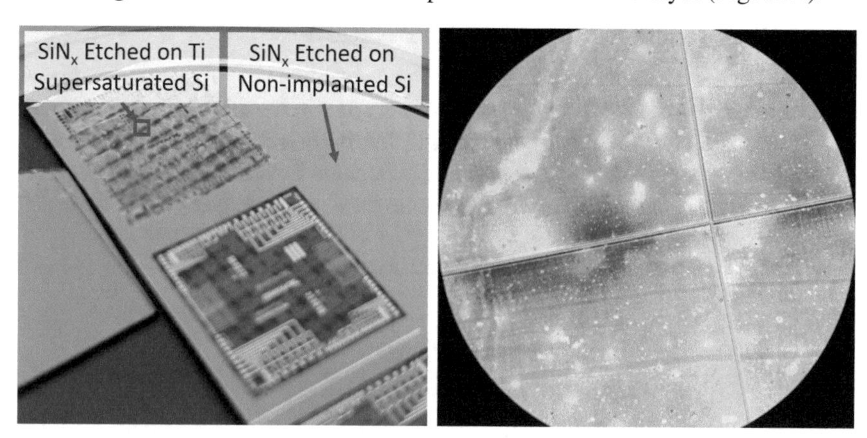

Fig. 3.24 Left: photograph of half a wafer showing three different areas: the square-shaped area in blue colour, where silicon nitride was wet etched over a Ti supersaturated layer; the square-shaped area in purple colour where there is a nitride layer atop the Ti supersaturated layer; and finally the rest of the wafer, were nitride was deposited atop non-implanted Si and later etched. The red square delimits the area examined by the optical microscope. Right: photograph taken through the optical microscope of the blue area, where silicon nitride was etched atop the Ti implanted layer. The NLA overlapping lines are visible

After the wet etching process using BOE, we observed a thin layer of different colours, from green to blue, present on the surface of the Ti supersaturated layer. We found this layer to be chemically inert, it did not react to any of the solvents we had in our laboratories. This layer only appeared on the Ti implanted areas, which led us to associate this phenomenon with the presence of Ti atoms in the sample. This effect was not observed on samples were silicon oxide, instead of nitride, was deposited. Therefore, we suspect this unexpected residual layer could be a compound of nitrogen and titanium. Searching on the literature, we found that titanium nitride (or tinite) is a hard and relatively inert ceramic compound, used as a coating to stiff mechanical tools, avoiding corrosion. It is electrically conductive, although it is a good IR reflector [33], an undesirable effect on IR photodetectors. Some authors were able to deposit TiN using ECR-CVD [34], the same technique we used to deposit the dielectric layers, although they used gaseous precursors containing Ti. Another work proposed the formation of TiN layers by depositing Ti atop a wafer, which was later annealed at temperatures around 400°C in a nitrogen atmosphere [35]. Although more experiments should be done to identify the chemical composition and structure of this unexpected layer, it seems plausible that it could be formed during the ECR-CVD process, were high energy ions and moderate temperatures may be present in the surface of the wafer, where Ti concentration is very high.

It is important to note that this layer may be present under the metallic contacts on the Ti supersaturated layer, as we used the wet etching process to define the contact opening. In any case, we decided to switch to silicon dioxide as passivating layer instead of silicon nitride to avoid this layer in future devices.

3.7 Pixel Matrix Device Characterisation

The characterisation of the pixel matrix consists of two techniques: electrical characterisation through I-V measurements and the optoelectronic properties via the spectral photocurrent, which is later used to extract the External Quantum Efficiency (EQE). Both measurements are performed with the sample already placed in the chip carrier.

3.7.1 Current-Voltage Characteristics

After the manufacturing process, the first characterisation we perform on the fabricated devices is the current-voltage curve, using the four-probe station described in Section 3.2.4.3. Several dies were fabricated and measured. Here, we represent only a few representative devices, using a transversal configuration:

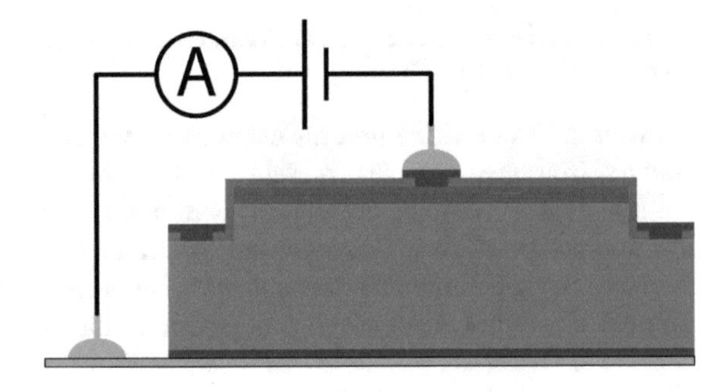

Fig. 3.25 Schematic cross-section of a pixel, measured using the transversal configuration

In Fig. 3.25, we represent a schematic of the IV measurement set up, although the Keithley 2636A is configured to measure using a three-wire configuration. The positive bias is applied on the substrate. Initially, we designed the guard ring in order to be used as one of the contacts for characterisation purposes, mainly, to increase the isolation between neighbouring pixels. The evolution of the Silfrared pixel matrix structure led us to design mesa devices in the last generation, as explained in Section 3.3.5. This decision made the guard ring to contact the bare silicon substrate, after the RIE process (see Fig. 3.25). Although aluminium is suitable to contact p-type silicon, depending also on the doping level at the surface, it may require thermal processes to produce ohmic contacts [36]. We have found experimentally that the vast majority of

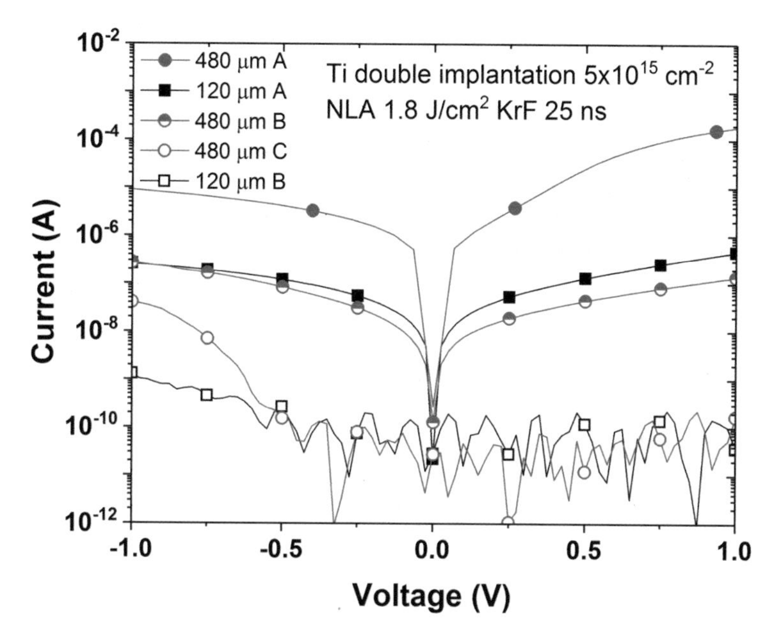

Fig. 3.26 Current-voltage curves of a few representative pixels of different sizes, double Ti implanted with a total dose of 5×10^{15} cm^{-2} and laser annealed at 1.8 J/cm^2

the guard ring contacts where rectifying, also featuring very low currents. Therefore, we discarded the use of guard ring contacts in the measurements performed within this thesis. Future designs may include ion implantation processes in the trench to enhance the isolation of pixels [37]. The increase on the doping level at the surface may ease the contact formation, even without thermal treatments [38].

Figure 3.26 shows five representative IV curves. We have experimentally found that most IV curves show three different behaviours. Around 80% of measured devices, measured in different dies of different wafers, displayed flat-like IV curves, with current values in the order of $10^{-8}-10^{-12}$ A for bias values between -1 and 1 V. In Fig. 3.26, this behaviour is represented by curves with open symbols. We will refer to this behaviour as the "insulating type". Around 15% of total devices displayed curves with resistive behaviour, featuring currents in the order of $10^{-6}-10^{-7}$ at 1 V. Finally, the remaining 5% of the fabricated pixels displayed a slight rectifying behaviour, with inverse currents in the order of 10^{-5} A at -1 V, and forward currents in the order of $10^{-4}-10^{-3}$ A at $+1$ V.

In order to establish a methodology to compare the different current-voltage curves of the measured devices, we calculated the resistance at $+1$ V and we found a soaring spread in the resistance values, up to seven orders of magnitude.

3.7.2 Quantum Efficiency

We evaluate the optoelectronic properties of the pixel using the QE. Most characterisation is performed in the IR cluster, although some pixels were also measured in the UV-Vis cluster. The vast majority of collected data was the short-circuit photocurrent, although we characterised the photocurrent at different bias for some samples.

Figure 3.27 depicts the QE curve in the sub-bandgap region for the same five pixels, whose IV curves were shown in Fig. 3.26. It seems there is a correlation between the maximum current in the IV with respect to the spectral photoresponse; the two pixels having the higher QE in the sub-bandgap region are also the pixels who displayed the highest direct currents. The five pixels shown in Fig. 3.27 displayed sub-bandgap response up to at least 0.7 eV, which corresponds to 1.74 μm at room temperature. However, for the two highest QE curves, the measurable photoresponse extends in our experimental system up to 0.45 eV, which is around 2.76 μm.

We analysed the spreading of the short-circuit photocurrent at 0.8 eV (1.55 μm) provided by different pixels and we found a similar spread in the photocurrent, around three orders of magnitude. In order to study the relationship between the electrical behaviour and the sub-bandgap photoresponse, we represent the short-circuit photocurrent at 0.8 eV as a function of the resistance at +1 V. Among all the fabricated dies (with 88 pixels per die), we represent the data of the four dies having the highest pixel performances, as evaluated by their rectifying behaviour and their photocurrent. The rest of the dies are not described here for the sake of clarity.

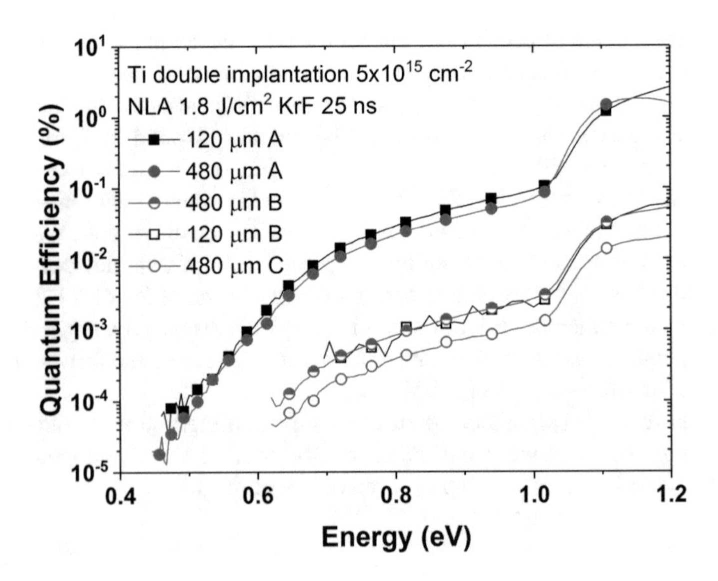

Fig. 3.27 Quantum Efficiency at room temperature of five representative pixels of different sizes, double Ti implanted with a total dose of $5\text{x}10^{15}$ cm^{-2} and laser annealed at 1.8 J/cm^2

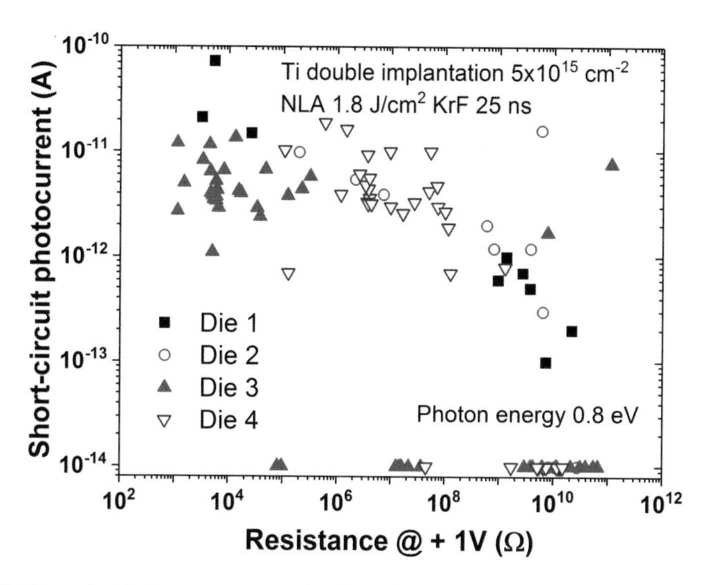

Fig. 3.28 Short-circuit photocurrent as a function of the resistance at +1 V for each pixel, when illuminated with photons at an energy of 0.8 eV

The majority of non-photo-responsive pixels exhibit high resistance values at + 1 V, between 10^9-10^{11} Ω. There may be a correlation between the resistance at + 1 V and the sub-bandgap response at 0.8 eV; pixels with lower resistance values display, in average, higher photocurrent than other pixels with higher resistance. We observe also that pixels belonging to the same die tend to be grouped in Fig. 3.28, in particular for dies 3 and 4, which, on the other side, are the dies showing the highest ratio of functioning pixels.

Next, we study the influence of the polarisation in the QE. We have seen in Fig. 3.26 that some pixels showed a rectifying behaviour, which were also the pixels exhibiting the best sub-bandgap quantum efficiencies. We chose one of the rectifying-like pixels, which we polarised in reverse by negatively biasing the substrate, while grounding its upper electrode (to contact the Ti supersaturated layer). We polarise the device in reverse aiming to increase the width of the space charge region of the pixel, which could result in a QE increase [39]. The electrical configuration would be similar to what was shown in Fig. 3.25, but instead of using an ammeter, we use the lock-in amplifier to measure the photocurrent.

Following results shown in Fig. 3.29, we see that QE is increased by more than one order of magnitude in the whole sub-bandgap range when bias is increased. However, it also increases the noise level in more than two orders of magnitude. This affects the minimum detectable energy of the photon: the pixel, while short-circuited, is responsive up to 0.45 eV, but when polarised, it is responsive only up to around 0.55 eV. Aiming to obtain more information on the effect of the polarisation, we calculated the QE at 0.8 eV as a function of the polarisation for several pixels (Fig. 3.30).

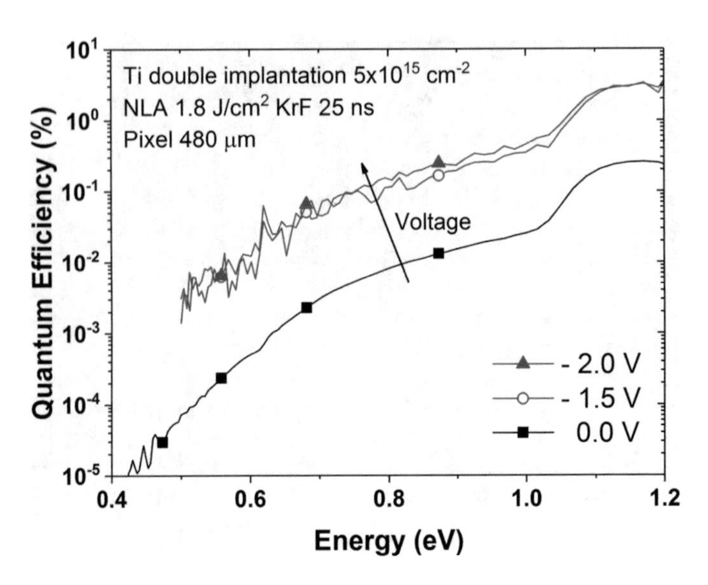

Fig. 3.29 Influence of device polarisation on the spectral quantum efficiency for a big pixel (480 μm). It was double Ti implanted at a total dose of $5x10^{15}$ cm^{-2}, laser annealed at 1.8 J/cm^2

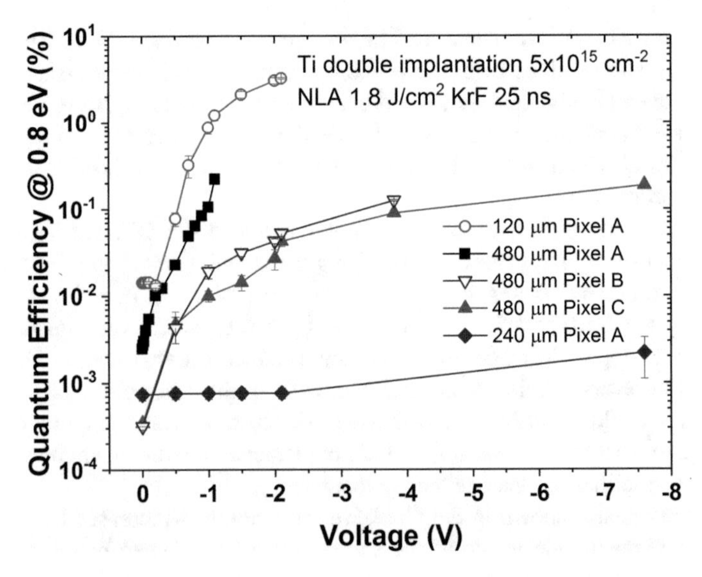

Fig. 3.30 Quantum efficiency at 0.8 eV as a function of the applied bias for several pixels, double Ti implanted at a total dose of $5x10^{15}$ cm^{-2}, laser annealed at 1.8 J/cm^2

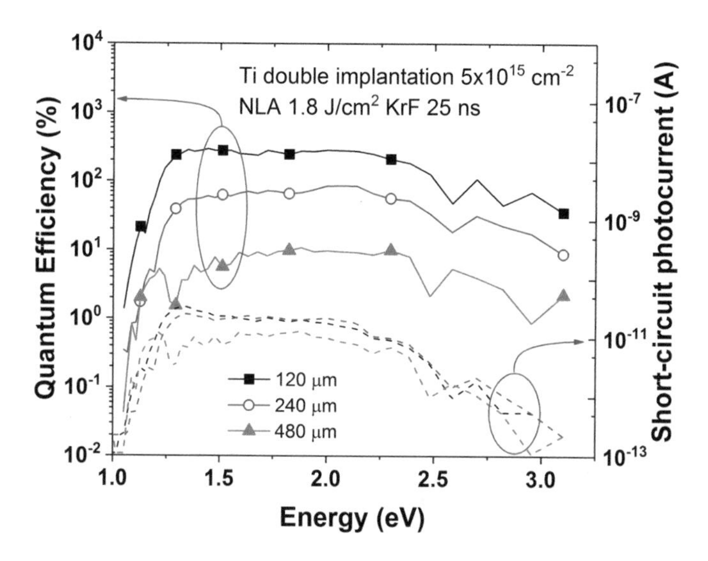

Fig. 3.31 Left axis: Quantum efficiency as a function of the photon energy in the UV and visible range for different pixels, double Ti implanted with a total dose of $5\text{x}10^{15}$ cm^{-2}, laser annealed at 1.8 J/cm^2. Right axis: measured spectral short-circuit photocurrent of the same pixels

The error bars indicate the average noise level of the measurement. There is a considerable increase in the QE when applying a positive bias on the Ti supersaturated layer (upper contact of the pixel), in some cases more than two orders of magnitude. As it happened to the rest of the measurements, there is a spread in the QE-voltage curve. Finally, we study the QE in the visible range at room temperature, using the UV-Vis cluster.

Figure 3.31 represents the QE curves of three pixels of different sizes on the left Y-axis, while the raw spectral short-circuit photocurrent is displayed on the right Y-axis. The most remarkable feature of Fig. 3.31 is that the QE is higher than 100% for the 120 μm pixel, while the others do not exceed the 100% mark. We include the raw spectral photocurrent in the right Y-axis to compare the signal provided by different pixels. The bigger pixel produced the lower photocurrent, while the medium and small pixels provided almost the same current. Their QE difference comes from the normalisation to their different pixel areas.

3.7.3 Discussion

The optoelectronic characterisation, by means of IV curves and photocurrent measurements, showed that there was a wide variability between pixels within the same pixel matrix, of several orders of magnitude. Besides, we observed that around 80% of the pixels were not functioning. Among the rest, only 5% showed rectification, which we related to higher sub-bandgap photoresponse.

QE measurements showed that microscale pixels, based on p-type Si substrates, double Ti implanted with a total dose of $5x10^{15}$ cm^{-2}, subsequently followed by a NLA process at 1.8 J/cm^2 exhibit sub-bandgap photoresponse up to 0.45 eV at room temperature. However, as it happened with electrical measurements, we found a high dispersion in the QE results.

After the analysis of the data coming from hundreds of pixels, we observed that the rectifying pixels were the ones producing the highest photocurrents in the sub-bandgap region. We could correlate a better QE with the IV characteristic; the pixels with a rectifying-like IV curve displayed the higher QE values in the sub-bandgap region also when reverse biased, similarly to what was shown in Fig. 3.28. We could find a QE at 0.8 eV as high as 3.1% at a reverse bias of –2.1 V on a small pixel, although the majority of analysed pixels exhibited QE values in the order of 0.2% at the same photon energy.

However, the exact QE values should be taken with care. Effectively, when we examined the QE in the visible range (Fig. 3.31), we observed that, in the case of the small pixel, we could find QE higher than 100%, up to 250%. It is physically impossible to obtain QE values higher than 100%, unless considering avalanche or amplifying effects, which are not expected to happen in our devices. Therefore, QE values over 100% could be related to cross-talk effects, in which neighbouring pixels are contributing to the photocurrent measured in one pixel. In most cases it is very difficult to obtain the real area contributing to the photocurrent when cross-talk is present. In order to have an estimation of this effect in the QE of the pixels, we simulated the pixel structure in 2D using Athena and Atlas simulators from the Silvaco TCAD suite. We used a N^+P diode, using the same dimensions and doping levels as in the real device. Instead of Ti, we used phosphorus in the ion implantation process. The energy was set to rend P profiles similar to those experimentally measured for Ti atoms. We performed several simulations under dark and illumination conditions to evaluate the cross-talk. In a first approach, we illuminated only the pixel labelled as 1 in inset of Fig. 3.32 with a monochromatic beam of variable energy. Then, we extracted the simulated photocurrent of each pixel and represented the three curves:

TCAD simulations, represented in Fig. 3.32, display that there may be crosstalk present in our devices, as a measurable photocurrent on the non-illuminated pixels. The illumination configuration, shown in the inset, is usually called the "underfilled" configuration, where part or the whole pixel is illuminated, in contrast to the overfilled configuration, in which not only the whole pixel is illuminated, but the surroundings as well. Our experimental set up measures the pixels in the overfilled configuration. In a second approach, we illuminated the whole structure with a fixed monochromatic beam set to 2.0 eV and measured the current in the middle pixel, the number 2. In these simulations, we vary the height of the mesa structure to see the effect of this parameter in the cross-talk.

There seems to be a critical value in which the QE drops down from 200% to around 86%. This value is dependent on the energy of the impinging photons and the intrinsic properties of the semiconductor as, e.g., the carrier lifetime or the dopant distribution in the mesa structure, among others. In Fig. 3.33, the critical height value of the mesa structure is 0.63 μm, below which cross-talk can be predominant on the

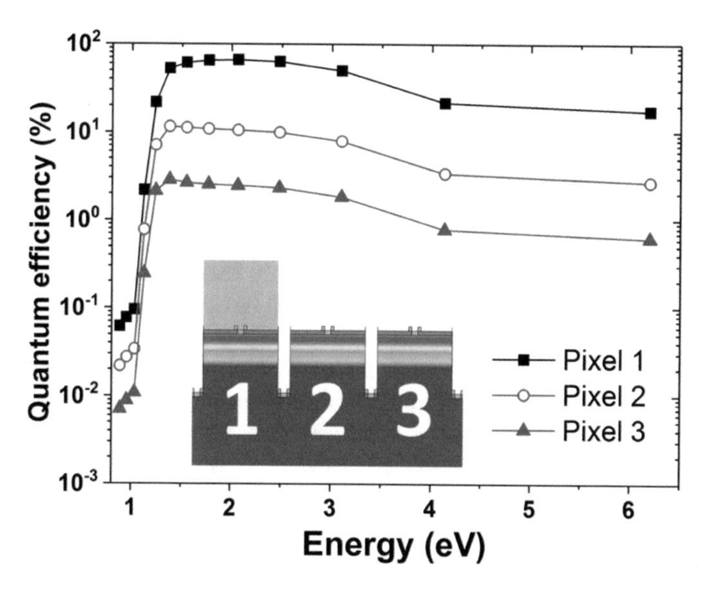

Fig. 3.32 Silvaco TCAD simulation of the quantum efficiency, calculated at each pixel when illuminating only pixel 1. The inset shows the simulated illumination configuration by a golden rectangle, where only pixel 1 was illuminated

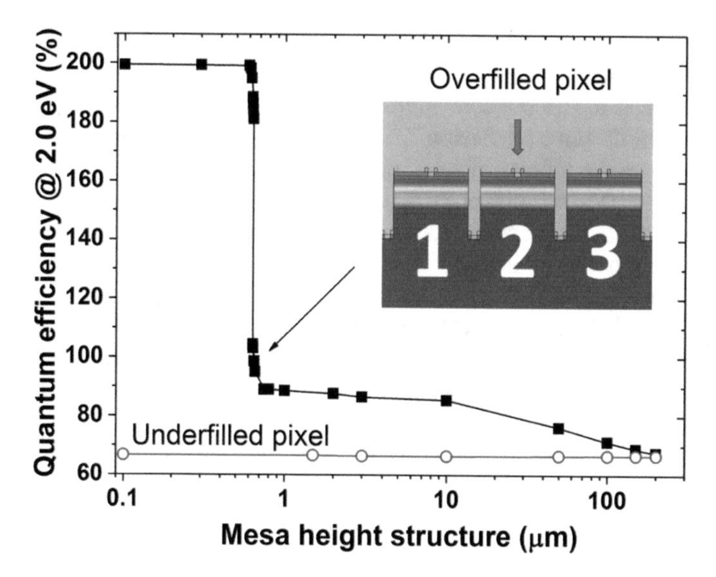

Fig. 3.33 Silvaco TCAD simulation of the quantum efficiency, at a fixed photon energy of 2.0 eV for different heights of the mesa structure. The open red circles represent the underfilled measurement of Fig. 3.32. Inset shows the overfilled illumination configuration. Photocurrent of pixel 2 was measured in this simulation

photocurrent measurements. However, even for values higher than the critical value, cross-talk is still affecting the pixel; from the overfilled QE of around 86% to the underfilled QE of 66%. The extra 20% comes from the contribution of the other two pixels: the cross-talk. Thus, even after the critical value, cross-talk is still present, up to around 200 μm, where both curves meet. This is why most commercial pixels nowadays use chemical-mechanical polishing techniques to thin the wafers down to a few microns, where they use trenches to isolate the pixels [37]. Therefore, we propose to increase the height of the mesa structure in future generations to reduce the cross-talk, as the Ti implanted pixels could still lie in the region of high crosstalk with the current height of 1.5 μm.

Until what extent could cross-talk be applicable to sub-bandgap photons? In visible and NIR light, all the wafer is capable to absorb photons, contributing to the photocurrent. However, photons with energy lower than the bandgap can only be absorbed in the Ti supersaturated layer, which is only located atop the device. Therefore, the crosstalk in the sub-bandgap region could be of second order, as the only way the photogenerated carriers would have to contribute to the conduction mechanisms would be to diffuse from the upper part of the neighbouring pixels. We expect to perform simulations increasing the sub-bandgap absorption coefficient of the pixel structure to have an estimation of the cross-talk.

Our current experimental configuration does not allow us to illuminate pixels individually (the underfilled configuration), as we did in TCAD simulations. It could be done using a focusing set-up. However, focusing an infrared beam in such small area with enough accuracy would be a challenging task and would require special lenses and IR detection equipment, none of which is available at our facilities. Therefore, until further revision we will consider the experimental QE values as an estimation and not exact values.

Coming back to the possible source of the inhomogeneity, we used the FIB-SEM technique described in Secton 3.2.3.2 to find out the differences between functioning and non-functioning pixels, by observing their structure in cross-section. Cross-section FIB-SEM proved to be particularly useful when identifying possible fabrication defects. In particular, we found a steep mesa profile, abrupt enough to possibly interrupt the continuity of deposited layers, as previously seen with nitride layers (refer to Fig. 3.22). The same could be happening to the metallic lines.

Figure 3.34 A displays an overview of the metallic track going from the mesa structure (upper half) to the pit surrounding the pixels (middle part), towards the metallic bonding pad. Note that the lower part of the micrograph (close to the legend) is located at the same height as the mesa structure, as previously shown by optical profilometry in Fig. 3.20 left. Thus, each metallic line has to climb up to two different hills, which should have similar profiles. From the detailed micrograph in Fig. 3.34 B we cannot assure the continuity in the metallic line, compromising the transmission of the electrical signal coming from the Ti supersaturated layer atop the mesa-structured pixel.

We suspect that the fabrication processes, in particular the RIE process, could severely influence the electrical properties of the pixel matrices. The mesa height profile may be too steep, abrupt enough to compromise the continuity of deposited

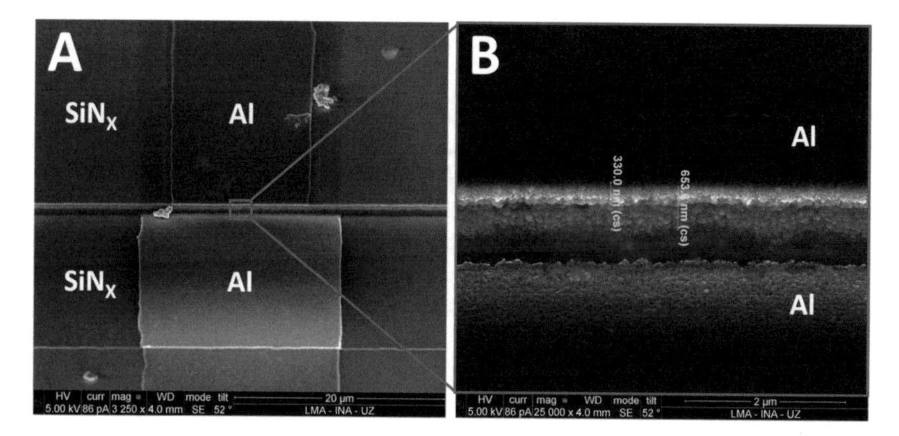

Fig. 3.34 SEM micrographs of a metallic line obtained at a tilt angle of 52°, corresponding to the pixel upper electrode. (**A**) Overview. (**B**) Detailed micrograph of the red boxed area in micrograph A

layers. For dielectric layers, it could compromise the reverse saturation current and the shunt resistance values. The case of metallic layers is more detrimental to the device performance: one broken line would render the pixel useless. Metallic tracks partially broken could also result in an increased series resistance.

3.7.4 Conclusions on Microscale Devices

The main purpose of fabricating microscale prototypes is to study the viability and suitability of the Ti supersaturated material for further integration into a possible commercial device, aiming to extend Si responsivity at energies lower than the bandgap. As starting material, we used p-type Si substrates, Ti implanted with a double ion implantation process, with doses of 10^{15} and $4x10^{15}$ cm^{-2} at an energy of 35 and 150 keV, respectively. After, we used a NLA process to recover the crystal quality of the implanted layer, by means a KrF laser with a pulse duration of 25 ns, at a fluence of 1.8 J/cm^2. The use of a p-type substrate is a novelty in our research group, as our previously published works relied on n-type Si substrates.

The characterisation process resulted in around only 5% functioning devices, which we attributed to a fabrication problem in the definition in the mesa structures, that compromised the continuity of the metallic lines. The functioning microscale pixel prototypes allowed us to demonstrate a proof of concept, in which we extended the responsivity of Si devices in the SWIR region of the spectra, up to 0.45 eV at room temperature, way below the bandgap of Si. The material may have the potential to be integrated in a commercial device, although our initial premise of homogeneity could not be tested due to the already described fabrication problems. We expect to optimise the fabrication route, which could increase the rate of functioning pixels, as well as improving their optoelectronic properties, mainly the QE.

3.8 Improvements on the Fabrication Route

Following the results shown in the previous sections, we decided to stop the pixel matrix fabrication process using Ti supersaturated layers in order to focus on the problems that arose during the fabrication and characterisation processes. In particular, we focused on two main problems: the dry etching process (RIE) and the metallisation deposition. The first process reported problems during the ulterior deposition processes of the dielectric and metallic layers, while the second process could affect the electrode reliability when contacting the device in both Ti implanted and non-implanted silicon.

3.8.1 Contact Optimisation

After the fabrication of the first prototypes, we realised that our contact fabrication technology ought to be improved for transversal device fabrication. Until that point, we had been using a combination of 50 nm of Ti followed by a deposition of 100 nm of Al by e-beam evaporation to the vast majority of the fabricated samples. We used the Ti layer in contact with silicon because of its oxygen scavenging effect on the native oxide, easing the direct contact between the metal and the semiconductor. The aluminium layer would act as an antioxidaion barrier for Ti and as the metal in contact with the characterisation equipment. Most of the material characterisation we were working on, before the development of this thesis, was based on what we have called "horizontal" characterisation, by means of the van der Pauw configuration. As we published before [40], these methods are relatively strong with respect to the possible effect of the contact quality. Even poor contacts could still be used in four-probe measurements. However, it is not the case when doing transversal measurements.

To further study the effect of the contacts, we fabricated N^+P and P^+N diodes following the same fabrication route we did with the Ti implanted substrates. We started with one wafer of each P2 and N1 substrates (Table 2.1), which are basically identical, except for their different type. Both wafers received the following ion implantation process:

- Boron implantation. Dose 10^{15} cm^{-2}, energy 35 keV. On frontside of n-type wafer and on backside of p-type wafer (to act as BSF).
- Phosphorus implantation. Dose 10^{15} cm^{-2}, energy 80 keV. On frontside of p-type wafer and on backside of n-type wafer (to act as BSF).

After, both wafers were submitted to an RTA process of 1000°C during 20 s in Ar ambient to activate the dopants on both sides of the wafer. The wafers followed the recipe of the sixth Silfrared generation, and we used the same metallisation on both surfaces. After the fabrication process we took the whole wafers to the probe station to measure the current-voltage characteristics. As it happened to Ti implanted wafers, most pixels did not work, possibly due to the RIE process, which was not modified at the moment of fabrication of these samples. After the first measurement process, we performed a FGA process on both wafers at 300°C during 20 min. We measured again the IV curves of the same functioning devices. Later we performed another FGA at 400°C during 20 min, and we measured them again. We simulated the structures using the Athena module of the Silvaco TCAD Suite to reproduce the implantation, RTA, etching and deposition processes with the same experimental values (doses, energies, times, temperatures, thicknesses). After the structure is defined, we use the simulator Atlas to define the electrical parameters and models of the simulation. The structure is similar to what was shown in inset of Fig. 3.32.

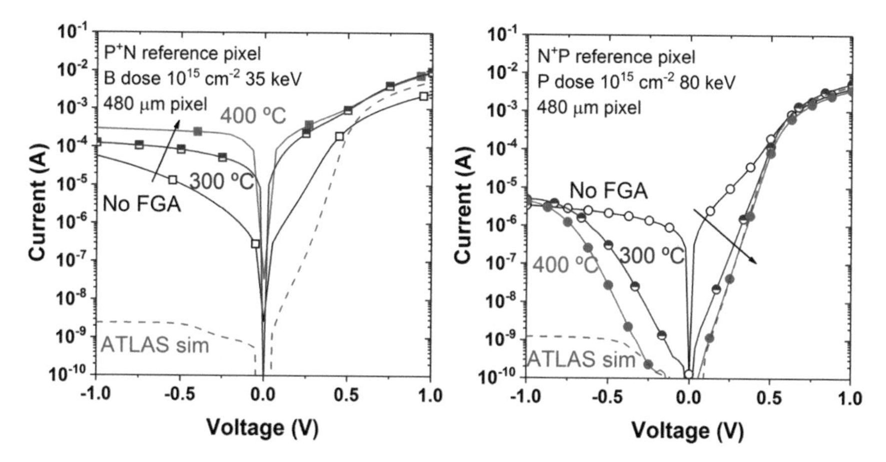

Fig. 3.35 Experimental and simulated IV curves for a P+N big pixel (left) and a N+P big pixel (right). Simulations were performed using Silvaco Athena + Atlas simulators with the real fabrication parameters. In the left figure, positive bias was applied to the pixel, while in the right figure it was applied to the substrate, so that forward bias is represented always on the right

The reference diodes did not show the expected rectifying behaviour before the FGA processes. After the FGA at 400°C, the N+P diode showed a direct current almost identical to the simulation, indicating that the contacts were close to ideal (as in the simulations, where they were simulated as ohmic). However, in the case of the P+N diode the rectifying behaviour was lost after the FGA processes. The reason behind the different behaviour after the thermal treatment of both substrate types could lie in the different ion implantation process that led to the junction formation. In n-type substrates, we used a boron implantation at only 35 keV, while in p-type substrates we used a phosphorus implantation at 80 keV. The metallurgical junction (the point in which the concentration of both types of dopants is the same) as simulated by TCAD was 460 nm for P+N pixels, and 640 nm for N+P pixels. Aluminium spiking could be present in both samples, but since the junction is deeper in the case of N+P pixels, they could handle better the thermal treatments.

The fabrication of PN diodes, together with TCAD simulations allowed us to check that the pixel fabrication route could end up with functional photodiodes.

However, thermal annealing processes were necessary in both cases, even though both surfaces were ion implanted with relatively high doses. This result would point out to a possible deficient back surface contact on our Ti implanted pixels, as the ion implantation used in the BSF on p-type substrates was the same. Following this experiment, we introduced two goals to achieve before continuing the device fabrication (Fig. 3.35):

- BSF doping processes ought to be changed, namely to higher doses and energies, to assure better contacts, as suggested in some studies [41].
- Ohmic contacts without any thermal treatment, to avoid a possible degradation of the Ti supersaturated material.

On line with the first decision, we performed a series of simulations, aiming to reach a dopant concentration value at the surface of 6×10^{19} cm^{-3}, which would be an adequate concentration value to obtain a tunnel-assisted ohmic contact, as described in reference [41]. The improved ion implantation parameters are:

- First ion implantation process of B or P at 35 keV and 5×10^{15} cm^{-2}.
- Second ion implantation process of B or P at 150 keV and 5×10^{15} cm^{-2}.
- RTA process at 1100°C and 120 s.

After the feedback coming from the fabricated reference pixels, based solely on P and B ion implantation processes, we designed a new set of samples, aiming to find a suitable metal stack combination which would render ohmic contacts, ideally without any thermal treatment. We used the same substrates of the previous sub-section and implanted them with P and B respectively in both sides, with the recipes shown a few lines before. After, we deposited different metallic electrodes in the front and backside of the wafer and measured the IV curves, before and after FGA process at increasing temperature. At each sample, we deposited the same metallisation on both surfaces, aiming to have the same metal-semiconductor junction on both electrodes. After a literature review, we started to consider the sputtering technique as a possible solution to our ohmic contact problem [41–43].

Due to our fluid collaboration with the research group "Grupo de Investigaciones Fotónicas" (GRIFO) of the Alcalá de Henares University (UAH) we had access to their sputtering system, which was used to deposit Al layers on this set of samples. We used the next processes for metallization tests:

- 100 nm Al deposition by e-beam evaporation.
- 50 nm Ti + 100 nm Al deposition by e-beam evaporation.
- 100 nm Al deposition by RF sputtering (UAH facilities), after Ar sputtering on the substrate to remove the native oxide.

We used electrodes with the same size ($50 \times 50\ \mu m^2$) to compare between different samples. The experimental results point to the 100 nm Al electrode deposited by RF sputtering as the best option for both N^+ and P^+ type substrates, before and after the FGA process. We observe how the FGA affects little the IV curves, as opposed to what we observed in the reference diodes (Fig. 3.36). Designing and mounting the

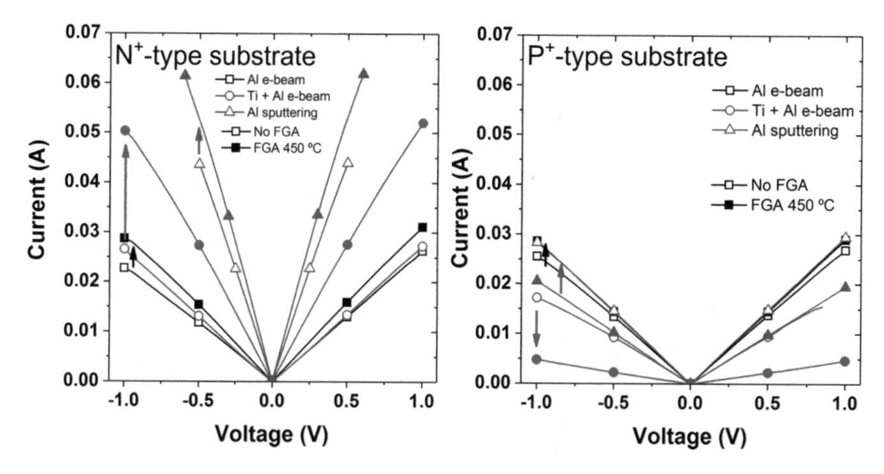

Fig. 3.36 Current-voltage characteristics obtained for an n-type (left) and p-type (right) implanted on both sides with P and B with a total dose of 10^{16} cm^{-2}, respectively. They were later deposited on both sides with Al by e-beam evaporation (black), Ti + Al by e-beam evaporation (red) and Al by RF sputtering (green). Open symbols stand for as-deposited, while solid symbols represent the samples after a FGA process of 450 °C during 3 min. Coloured arrows indicate the change of the IV curve after the FGA process

homemade sputtering system has been part of this thesis, aiming to provide a new metal deposition technique to our research group that could solve the metallisation problems described in this sub-section.

Given these results, and considering that the fabricated Ti supersaturated Si dies were not submitted to any FGA process, there could be a strong limiting factor on the contact properties on both the Ti implanted layer and the substrate. Therefore, depositing contacts with better electrical properties could be the key to further improve the sub-bandgap photoresponse on the microscale devices.

3.8.2 Dry Etching Optimisation

The second biggest issue we found on our first prototypes was the steepness of the fabricated mesa structures, which showed to be problematic at the moment of depositing layers atop them. In this sub-section, we delve into the dry etching process, in search for a smoother etched profile for future device fabrication.

Reactive ion etching technique is relatively flexible with respect to the degree of anisotropy that can be achieved during the etching process. As mentioned in Section 3.2.5.5, the etch rate and profile are strongly dependent on the pressure and RF power during the process. Low pressure and high RF power processes would tend to be anisotropic, producing step-like structures, while high pressure and low RF power would produce isotropic structures, characterised by rounded, smoother step profiles.

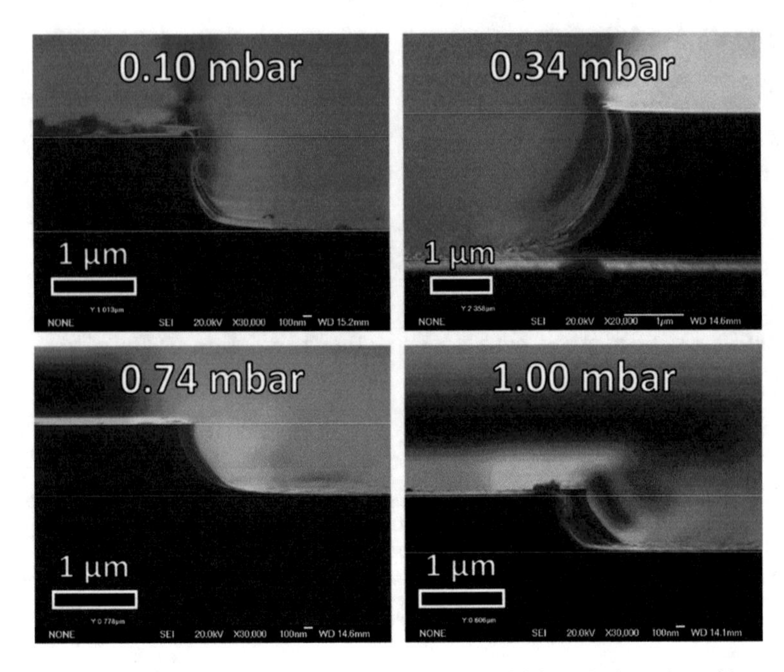

Fig. 3.37 Cross-section SEM micrographs of mesa structures fabricated using a dry etching process under several SF_6 pressure conditions, at a fixed RF power of 10 W during 4 min

From the SEM micrographs shown in the previous section, we found out that the current configuration of the RIE process led to the formation of anisotropic-like structures, with walls nearly vertical. The step-like profile affected the integrity of deposited layers, which we want to avoid for future devices. The etched profile may indicate our previous RIE process was performed at low pressures or high RF power. The RF power was set to 10 W, already a low RF power value, which led us to increase the gas pressure in order to search for smoother profiles.

Therefore, we designed and fabricated a set of samples with increasing gas pressure, using the same substrate used for device fabrication. We defined the mesa structures prior the etching process, using a lithography process aligned with the flat of the wafer, followed by an aluminium deposition. We then cleaved the samples using a diamond point tool and examined the cross-section using SEM (Fig. 3.37).

Increasing the pressure effectively reduced the aspect ratio, from almost 6 at the pressure used for device fabrication (0.1 mbar) to 1.4 at 0.86 mbar.

The goal of the RIE optimisation study was to avoid losing the continuity of the deposited layers. In order to check this hypothesis, we performed another lithography process over the mesa-structured sample, using a stripe-like mask. Then, we deposited a 200 nm thick SiO_2 layer, followed by a 100 nm thick aluminium layer (to simulate the real conditions of the pixels). Using a lift-off process, we defined straight metallic lines (thinner than the mesa structure), over several mesas. Using our probe station, we checked the continuity between the upper surface of the mesa device and the pit to see if the metallic line was cut at some point while crossing the etched profile. We measured that samples with SF_6 pressures lower than 0.34 mbar contained broken lines in an important percentage of the devices, quantity that diminished as the pressure increased. In the case of pressures equal or higher than 0.74 mbar we could measure continuity in all the examined dies. In fact, in those samples, we could measure continuity between two points of a metallic segment containing 13 mesas in between, i.e. up to 24 different steps were crossed by the metallic line, and it still had continuity.

Therefore, we could experimentally demonstrate that etching the mesa structures at pressures equal or higher than 0.74 mbar, at an RF power of 10 W, could solve the aforementioned problem of continuity in the deposited layers.

3.9 Conclusions

In the first sections of this chapter, we described the manufacturing and characterisation processes of microscales devices based on Ti supersaturated Si, aiming to extend silicon responsivity towards the sub-bandgap region, the SWIR and NIR region, at room temperature.

In Section 3.3.3, we discussed the possible origin of the sub-bandgap response by analysing the effect of a SiO_2 passivation layer atop the implanted layer, comparing the results with a non-implanted sample. The results showed that the sub-bandgap response of the Ti implanted sample increased by a factor of 30, while the reference sample diminished its responsivity around one order of magnitude, possibly due to passivated surface defects. The responsivity value found on this thesis, after the oxide deposition process, improved our previously reported highest responsivity value in a factor of almost 20, up to 590 mV/W at 0.8 eV at room temperature.

In Section 3.3.4, we analysed the possible influence of some inhomogeneity sources found on the fabricated Ti supersaturated material. Using several techniques, we could not measure the effect of the laser overlapping lines on the pixel structure nor the electrical properties. Another important inhomogeneity issue are the craters, present on the surface of all the samples. From EDX measurements, we proposed that craters could be the result of an interaction with dust or solid particles present on the surface during the laser process. Finally, we analyse the cellular breakdown effect, visible on the Ti supersaturated layers used in this sub-section. We do not expect variability sources derived from the CBD structures due to their relatively small size, when compared to the pixel area.

After, we described the pixel matrix fabrication and characterisation steps. First, we designed the pixel structure, which led to the design of a set of photolithography masks. The fabrication process involved more than 100 steps to obtain the final device, involving several techniques described in Section 3.2.5. During this thesis, up to six different generations of devices were fabricated, whose complete manufacturing details can be found in the Annex B. During and after the fabrication process, we characterised the different steps to check that the manufacturing was done according to the expected specifications.

The electric and optoelectronic properties of the fabricated pixel matrices were obtained by means of current-voltage measurements for the former and spectral photocurrent for the latter. The results showed a wide spread on the collected data, on both the IV curves and the QE measurements. Around 80% of the fabricated pixels did not work and could not be measured. We identified the root cause of the inhomogeneity as a problem in the fabrication process, in particular the steepness of the mesa structure, which resulted in broken metallic lines, rendering the pixels useless. Of the functioning pixels, most of them exhibited sub-bandgap response at room temperature, demonstrating that it is possible to fabricate microscale devices responsive to sub-bandgap photons. We performed a statistical analysis on both the electric and optoelectronic properties, with which we could link low sub-bandgap photocurrent values to a poor electrical behaviour. Later, we analysed the influence of the polarisation on the QE, where we measured values up to 800 times higher when reverse polarising the pixels, at the expense of increasing the noise level at a higher rate. Values as high as 3.1% QE at the key photon energy of 0.8 eV were measured, although most functioning devices displayed values around 0.2% at the same wavelength. In a last characterisation step, we measured the QE at energies higher that the bandgap, i.e., in the visible and NIR range up to 1.2 eV. Examining these measurements, we realised that there could be a cross-talk effect between pixels, at least for photons with energies higher than the bandgap, which was later confirmed by TCAD simulations. Based on the simulations, we decided to increase the height of the mesa structure, aiming to minimise the effect of the parasitic signal coming from the neighbouring pixels. However, the possible effect of the cross-talk in the sub-bandgap region is yet to be determined, as the IR-absorbing layer is located atop the pixel, close to the carrier collection electrode, relatively far from the other pixels.

In the last section, we analysed the possible failure causes of the previously fabricated devices, aiming to further increase the performance of the pixel matrices based on Ti supersaturated silicon. In order to further understand the pixel structure, we fabricated N^+P and P^+N pixel structures, without Ti ion implantation. We found that there was still a large inhomogeneity in the process, possibly caused by the steepness of the mesa structure, as analysed by FIB-SEM micrographs. P^+N photodiodes (boron implantation on n-type substrates) showed poor performance, even after a FGA process, possibly due to Al spiking through the junction. N^+P photodiodes showed improved properties after the FGA process. On the functioning pixels, we could experimentally find a relationship between the electrical properties and the size, which was later confirmed by TCAD simulations.

In order to further understand the electrical behaviour of both Ti implanted and PN (non Ti implanted) pixels, we fabricated samples with different metallisation layers aiming to check the linearity of the contacts between the deposited metal and the semiconductor. The results encouraged the use of Al as the metal of choice, although it should be deposited by RF sputtering, instead of e-beam evaporation as we had been doing in the past. Due to these results, we started the design and fabrication process of a homemade sputtering system, in order to achieve full autonomy on the fabrication process of the Ti supersaturated samples. All the Al RF sputtered layers in this chapter has been deposited thanks to the collaboration of the GRIFO research group from the UAH.

Finally, we characterised the RIE process of the mesa formation as a function of the SF_6 gas pressure, in order to find a smoother etched profile, with lower aspect ratios. The previously fabricated samples had an aspect ratio higher than 5 (gas pressure of 0.1 mbar). With gas pressures in the order of 0.86 mbar we could lower this value down to around 1.5, which resulted in better deposited layers, as confirmed by electrical measurements on deposited metal stripes atop several mesa structures.

References

1. Dhariwal SR, Kulshreshtha AP (1981) *Theory of back surface field silicon solar cells.* Solid State Electron 24:1161–1165
2. Olea J, Pastor D, Martil I,Gonzalez-Diaz G (2010) *Thermal stability of intermediate band behavior in Ti implanted Si.* Sol Energ Mat Sol C 94:1907–1911
3. Wang M. et al. (2019) *Thermal stability of Te-hyperdoped Si: Atomic-scale correlation of the structural, electrical and optical properties.* Physical Review Materials 3, 044606
4. Newman BK, Sher M-J, Mazur E, Buonassisi T (2011) *Reactivation of sub-bandgap absorption in chalcogen-hyperdoped silicon.* App Phys Let 98:251905
5. Olea J, Toledano-Luque M, Pastor D, Gonzalez-Diaz G, Martil I (2008) *Titanium doped silicon layers with very high concentration.* J Appl Phys 104
6. Gonzalez-Diaz G et al. (2009) *Intermediate band mobility in heavily titanium-doped silicon layers.* Sol Energ Mat Sol C 93:1668–1673
7. Olea J, Gonzalez-Diaz G, Pastor D, Martil I (2009) *Electronic transport properties of Ti-impurity band in Si.* J Phys D Appl Phys 42
8. Antolin E et al. (2009) *Lifetime recovery in ultrahighly titanium-doped silicon for the implementation of an intermediate band material.* Appl Phys Lett 94
9. Olea J et al. (2009) *High quality ti-implanted si layers above solid solubility limit.* Proceedings of the 2009 Spanish conference on electron devices, 38–41
10. Olea J et al. (2010) *High quality Ti-implanted Si layers above the Mott limit.* J Appl Phys 107
11. Olea J, Pastor D, Toledano-Luque M, Martil I, Gonzalez-Diaz G (2011) *Depth profile study of Ti implanted Si at very high doses.* J Appl Phys 110
12. Olea J, del Prado A, Pastor D, Martil I, Gonzalez-Diaz G (2011) *Sub-bandgap absorption in Ti implanted Si over the Mott limit.* J Appl Phys 109
13. Olea J et al. (2011) *Two-layer Hall effect model for intermediate band Ti-implanted silicon.* J Appl Phys 109
14. Pastor D et al. (2012) *Insulator to metallic transition due to intermediate band formation in Ti-implanted silicon.* Sol Energ Mat Sol C 104:159–164
15. Garcia-Hemme E et al. (2012) *Ion implantation and pulsed laser Melting processing for the development of an intermediate band material.* AIP Conf Proc 1496, 54–57
16. Garcia-Hemme E et al. (2012) *Sub-bandgap spectral photo-response analysis of Ti supersaturated Si.* Appl Phys Lett 101
17. Olea J et al. (2012) *Low temperature intermediate band metallic behavior in Ti implanted Si.* Thin Solid Films 520:6614–6618
18. Pastor D et al. (2012) *Interstitial Ti for intermediate band formation in Ti-supersaturated silicon.* J Appl Phys 112
19. Garcia-Hemme E et al. (2013) *Double ion implantation and pulsed laser melting Processes for third generation solar cells.* Int J Photoenergy

20. Garcia-Hemme E. et al (2013) *Electrical properties of silicon supersaturated with titanium or vanadium for intermediate band material*. Proceedings of the 2013 Spanish conference on electron devices (CDE 2013), 377–380
21. Olea J et al. (2013) *Ruling out the impact of defects on the below band gap photoconductivity of Ti supersaturated Si*. J Appl Phys 114
22. Garcia-Hemme, E. et al. *Far infrared photoconductivity in a silicon based material: Vanadium supersaturated silicon*. Appl Phys Lett 103 (2013)
23. Garcia-Hemme E et al. (2014) *Room-temperature operation of a titanium supersaturated silicon-based infrared photodetector*. Appl Phys Lett 104
24. Garcia-Hemme E. et al. (2015) *Meyer Neldel rule application to silicon supersaturated with transition metals*. J Phys D Appl Phys 48
25. Chiarotti G, Nannarone S, Pastore R, Chiaradia P (1971) *Optical absorption of surface states in ultrahigh vacuum cleaved (111) surfaces of Ge and Si*. Phys Rev B-Solid St 4, 3398
26. Kerr MJ, Schmidt J, Cuevas A, Bultman JH (2001) *Surface recombination velocity of phosphorus-diffused silicon solar cell emitters passivated with plasma enhanced chemical vapor deposited silicon nitride and thermal silicon oxide*. 89:3821–3826
27. Garcia-Hemme E (2015) *Respuesta infrarroja en silicio mediante implantación iónica de metales de transición*. Thesis dissertation
28. Ebong A, Chen N (2012) *Metallization of crystalline silicon solar cells: A review. high capacity optical networks and emerging/enabling technologies*, pp 102–109
29. Montero D (2015) *Elaboración de una ruta de fabricación de absorbedores de radiación IR basados en silicio*. Master Dissertation
30. Grover S (2016) *Effect of transmission line measurement (TLM) geometry on specific contact resistivity determination*. PhD Dissertation
31. Pereira MA, Diniz JA, Doi I, Swart JWJA (2003) *Silicon nitride deposited by ECR–CVD at room temperature for LOCOS*. Isolation technology 212:388–392
32. Kitamura R, Pilon L, Jonasz M (2007) *Optical constants of silica glass from extreme ultraviolet to far infrared at near room temperature*. Appl Optics 46:8118–8133
33. Roux L, Hanus J, Francois JC, Sigrist M (1982) *The optical properties of titanium nitrides and carbides: spectral selectivity and photothermal conversion of solar energy*. Solar Energy Materials 7:299–312
34. Boumerzoug M, Pang Z, Boudreau M, Mascher P,Simmons JG (1995) *Room temperature electron cyclotron resonance chemical vapor deposition of high quality TiN*. Appl Phys Lett 66:302–304
35. Nulman J (Google Patents, 1993) *Formation of titanium nitride on semiconductor wafer by reaction of titanium with nitrogen-bearing gas in an integrated processing system*. European patent no. EP0452921A2
36. Ting CY, Crowder BL (1982) *Electrical-properties of Al/Ti contact metallurgy for vlsi application*. J Electrochem Soc 129:2590–2594
37. Arnaud T et al. (2011) *Pixel-to-pixel isolation by deep trench technology: application to CMOS image sensor*. IISW Conference at Hokkaido, Japan
38. Nowicki RS (1983) *Electrical-properties of Al/Ti contact metallurgy for Vlsi applications*. J Electrochem Soc 130:1446–1446
39. Zalewski EF, Duda CR (1983) *Silicon photodiode device with 100% external quantum efficiency*. Appl Optics 22:2867
40. Gonzalez-Diaz G et al. (2017) *A robust method to determine the contact resistance using the van der Pauw set up*. Measurement 98:151–158
41. Saraswat K (2005) *Ohmic contacts*. Stanford University Advanced Integrated Circuit Fabrication processes, 25
42. Gambino JP, Colgan EG (1998) *Silicides and ohmic contacts*. Mater Chem Phys 52:99–146
43. Ting CY, Wittmer M (1982) *The use of titanium-based contact barrier layers in silicon technology*. Thin Solid Films 96:327–345

Chapter 4
Results: NLA Using a Long Pulse Duration XeCl Laser

4.1 Introduction

This chapter describes the research done on samples Ti implanted in our facilities at UCM, but laser annealed with the XeCl excimer laser described in Sect. 2.2.1, belonging to SCREEN-LASSE (Paris). This laser features a wavelength of 308 nm, a pulse duration of 150 ns at FWHM and a laser spot of $10 \times 10\ mm^2$, which allows for full die exposure processes. Among the reasons that motivated our change to this laser we highlight the following:

- Laser melting research line on IPG Photonics closed. Their laser was no longer available.
- We wanted a European partner to reduce the processing and shipping times.
- Full die exposure: avoiding overlapping lines was a desirable feature, in order to reduce possible inhomogeneities.
- Longer pulse duration, to study its influence on the crystal quality and the influence on cellular breakdown structures.
- 300 mm wafer handling capability, compatible with STMicroelectronics technology, necessary for future integration in a commercial pixel route.

The collaboration with STMicroelectronics (Crolles) started with a six-month internship, where the goal was to integrate the Ti supersaturated material into a commercial fabrication route using 300 mm wafers. Thus, it was necessary to perform a first characterisation process with the new laser. The results with these samples may determine the Ti ion implantation parameters and NLA processes chosen to be integrated within the STMicroelectronics technology.

D. Montero Álvarez, *Near Infrared Detectors Based on Silicon Supersaturated with Transition Metals*, Springer Theses,
https://doi.org/10.1007/978-3-030-63826-9_4

4.2 Material Fabrication

We fabricated a total of 18 wafers, with different parameters:

Substrate: n-type substrate (code N1 on Table 2.1 or p-type substrate (code P2).

Ti ion implantation process. We implanted three different ^{48}Ti doses at 5×10^{14} cm^{-2}, 5×10^{15} cm^{-2} and 2×10^{16} cm^{-2} in two steps: 20% of total dose at 35 keV and 80% at 150 keV.

The laser fluence range was chosen so it could potentially melt the whole implanted layer. In our research group, we used laser fluences between 0.6 and 1.8 J/cm^2 when using the KrF laser with a 25 ns pulse to fabricate Ti supersaturated layers. Several works, based on simulations and on experimental data, studied the influence of the laser pulse duration on the melting process of crystalline and amorphous silicon [1, 2]. They found that, when comparing laser processes done under different pulse durations, the fluence value is not as important as it is the incident power per cm^2, which can be obtained by dividing the laser fluence to the pulse duration. Ishihara et al. found a relationship between the onset of Si melting and the pulse laser duration, in which the threshold for c-Si and a-Si melting processes depends on the square root of the pulse duration [2]. For example, considering the fitting parameters used in the reference in the case of c-Si, the laser fluence interval of 0.6–1.8 J/cm^2 at a pulse duration of 25 ns would be equivalent to an interval of 1.1–3.3 J/cm^2 with a longer pulse duration of 155 ns. Therefore:

Laser fluence. We chose six different laser energy densities, from 1.0 to 4.0 J/cm^2, with an interval of 0.6 J/cm^2 between different samples.

From the 18 wafers used for this lot, nine were fabricated using N-type substrates, which were ion implanted on their backside using P, with a dose of 10^{15} cm^{-2} and 80 keV of energy, to create a BSF. The other nine wafers were fabricated on P-type substrates, back ion implanted with B at a dose of 10^{15} cm^{-2} and an energy of 35 keV to act as BSF. Lately, the implanted dopants of the 18 wafers were activated using a RTA of 1000 °C during 20 s, using Ar. From each group composed of nine wafers, three received the same Ti dose for a total of three different doses, following the Ti implantation recipes described before. Later, we cut the wafers into quarters; each quarter received a total of four 1 cm^2 square-shaped laser shots at the same laser fluence, with an overlapping between laser exposed areas of around 100 μm wide (Fig. 4.1).

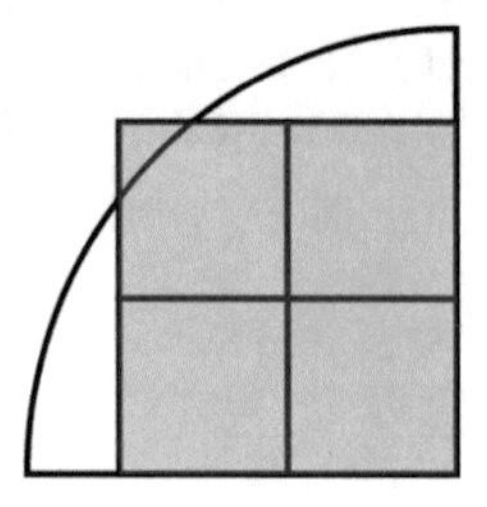

Fig. 4.1 Schematic of laser pulse (light red) positioning on a quarter of wafer. Pulses and wafer size are at scale

For device fabrication, we cut square-shaped samples of 9×9 mm^2, centred with respect to the laser shot to avoid the possible influence of the borders on the material properties (laser exposed area is 10×10 mm^2).

4.3 Material Characterisation

Prior to device fabrication, we must characterise the base material using different techniques, from the structural to electro-optical characterisation.

4.3.1 Structural and Compositional Characterisation

Ti profile distribution: SIMS

One of the first measurements performed on Ti implanted samples is the Ti distribution profile as a function of the depth. We used the ToF-SIMS equipment available at Universidad de Extremadura, described in Sect. 2.2.1, to experimentally verify if the Ti concentration was high enough to overcome the Mott limit.

We characterised eight different samples using this technique, all of them based on p-type substrates: three as-implanted samples, one per each different Ti dose, and five laser annealed samples. In the five laser annealed samples, we swept both the Ti dose and the laser fluence. We decided to take the middle Ti dose and sweep the laser fluence in a first set of three samples. Later, we took the middle laser fluence and swept the Ti dose (another three samples).

The as-implanted profiles feature two peaks, with are consistent with the double ion implantation process at two different energies. Examining the laser annealed samples, there is a strong Ti redistribution after the laser process, in which the implanted atoms tend to diffuse towards the surface, as previously reported by Olea et al. [3]. The sample with a dose of 5×10^{14} cm^{-2}, even though it did not achieve enough concentration to overcome the IB formation limit in the as-implanted case, finally exceeded it after the NLA process at 2.8 J/cm^2. We observe that the three annealed profiles overlap with the as-implanted profiles at increasing depths, when increasing the Ti dose.

In Fig. 4.2 right, we evaluate the influence of the laser fluence at a fixed Ti dose of 5×10^{15} cm^{-2}. After annealing the sample with a laser fluence of 1.6 J/cm^2, the Ti distribution profile did not change significantly, only in the first 100 nm, value after which it overlapped with the as-implanted curve. The sample annealed at a laser fluence of 2.8 J/cm^2 has been described in the previous lines. If we further increase

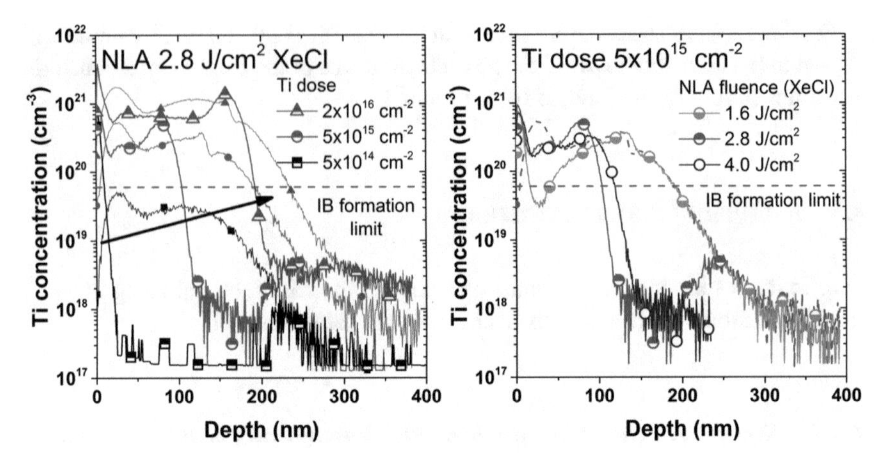

Fig. 4.2 Ti concentration profile in log Y-axis as a function of depth for three samples Ti implanted at different doses, laser annealed with a fluence of 2.8 J/cm^2 with the as-implanted references (left graph) and three samples Ti implanted at a dose of 5 × 10^{15} cm^{-2}, laser annealed at different fluences. As-implanted profiles are indicated in thinner lines with solid symbols for reference. The arrow indicates the direction of increasing Ti dose for the annealed samples (left)

the laser fluence up to 4.0 J/cm^2, the profile gets similar to the case of 2.8 J/cm^2, but with a slightly thicker supersaturated layer of 120 nm.

We can extract the thickness of the Ti supersaturated layer as the depth in which the concentration equals the IB formation limit (dashed red line in Fig. 4.2).

With respect to the samples Ti implanted at a dose of 5 × 10^{15} cm^{-2}, we did not include the supersaturated layer thickness of the sample with the lower laser fluence, due to its particular profile with a minimum around 40 nm deep. The retained Ti dose is included in the last column, where we observe that most of the Ti dose is retained after the NLA process.

Table 4.1 Ti supersaturated layer thickness as a function of the Ti dose and laser fluence using the XeCl excimer laser. The value of d_{TSL} represents the thickness of the Ti Supersaturated Layer (TSL)

Ti dose (cm^{-2})	Laser fluence (J/cm^2)	d_{TSL}(nm)	Overlapping with as-implanted (nm)	Retained Ti dose (%)
5 × 10^{14}	2.8	10	225	81.1
5 × 10^{15}	1.6	–	140	74.0
	2.8	103	250	69.9
	4.0	121	>230	64.6
2 × 10^{16}	2.8	192	280	75.9

Crystal quality: Raman spectroscopy

Once we have checked that the desired concentration of Ti has been reached in the samples, we want to characterise the crystal quality after the laser process. A first estimation can be obtained using Raman spectroscopy. We used the equipment available at the "CAI de Espectroscopía y Correlación", described in Sect. 2.2.4, at a wavelength of 532 nm. At this wavelength, the absorption depth is around one micron. However, it is possible to slightly adjust the depth by playing with the focusing system, so that only information coming from the most surficial volumetric region of the sample is sent to the spectrometer. The system was set to provide information of the shallower layer possible, estimated around 400 nm. We measured the Raman shift spectra as a function of the Ti dose and the laser fluence.

Figure 4.3 left shows the Raman shift of several samples with a Ti dose of 5×10^{14} cm^{-2}, laser annealed at different laser fluences, from the as-implanted sample up to 4.0 J/cm^2. We also measured a reference substrate. The reference spectra shows an abrupt peak at 521 cm^{-1}, related to the Transversal Optical (TO) phonon mode [4], followed by a flat shoulder on the right, up to 580 cm^{-1}. The as-implanted curve does not show any peak at 521 cm^{-1}. Instead, it features a broad band from (at least) 420 cm^{-1} up to 535 cm^{-1}, with a peak centred in 510 cm^{-1}. Another band appears, with negative slope, from 540 cm^{-1} up to the end of the measured spectra. We observe how the Raman spectra go from a curve similar to the as-implanted to a curve similar to the non-implanted reference as the laser fluence increases, expect for the peak, which is centred in 520 cm^{-1} instead of 521 cm^{-1}.

Figure 4.3 right show how the Ti dose influences the Raman shift spectra, at the two limit conditions: the as-implanted sample and those with the highest laser

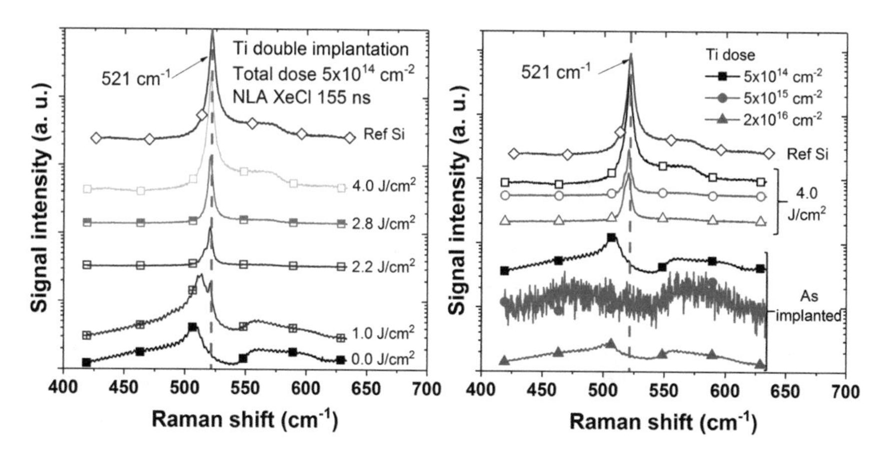

Fig. 4.3 Left: Raman shift spectra in log Y-axis of several samples Ti implanted with a total dose of 5×10^{14} cm^{-2} (grey scale), laser annealed at different laser fluences (the as-implanted sample is labelled as 0.0 J/cm^2), and a reference Si substrate (violet). Right: Raman shift spectra for sample Ti implanted at different doses, showing their respective as-implanted and laser annealed at 4.0 J/cm^2 profiles. In both graphs, the vertical dashed red line indicates the fundamental vibration mode of c-Si, at a Raman shift of 521 cm^{-1}

fluence. Dose increases from top to bottom on each group of three curves. The as-implanted samples show similar featureless curves. With respect to samples laser annealed at 4.0 J/cm^2, the fundamental peak shows an asymmetry to the left which becomes more pronounced with increasing Ti dose.

Crystal quality: TEM micrographs

In this section, we show the TEM micrographs of Ti implanted samples. We used the same five laser annealed samples that were analysed using SIMS. Besides the TEM micrograph, we use the Fast Fourier Transform (FFT) on HRTEM micrographs to obtain the diffraction pattern of the crystal structure, which can be used to estimate the crystal quality of the analysed surface. Finally, switching the microscope to STEM mode and with the help of the EDX technique we can estimate the chemical composition at a certain probing point of the sample. First, we analyse the sample with the lowest Ti dose, 5×10^{14} cm^{-2}, processed at a laser fluence of 2.8 J/cm^2.

Judging by the HRTEM micrographs, the crystal quality is very good, with no apparent difference between the Ti supersaturated layer and the rest of the Ti implanted and laser annealed area. The FFT images, which are a mathematical treatment of the labelled areas in the HRTEM micrograph, give an estimation of the Electron Diffraction pattern, related to the crystal quality [5] of the area. In both cases, inside the Ti supersaturated layer and the Si substrate (where no Ti atoms were detected by SIMS, only noise level), the pattern is formed by defined points, related to a monocrystalline structure [6]. Close to the Pt protective layer, we observe a clearer, amorphous-like layer around 4 nm thick.

In order to possibly identify the nature of this layer we must perform EDX measurements, which show that the amorphous layer close to the surface is composed mainly of silicon (66%) and oxygen (31%). The thickness, around 4 nm, points out to a native oxide layer. The exact numbers of composition must be taken with caution, as EDX is not a suitable technique to detect light elements as oxygen in a TEM lamella [7]. More advanced techniques as Auger spectroscopy will be performed in the future to further precise the composition [8, 9] (Fig. 4.4).

Now we move on to the samples with a total dose of 5×10^{15} cm^{-2}. At this dose, we performed the study as a function of the laser fluence. We start showing the results of the lowest fluence: 1.6 J/cm^2.

The left TEM micrograph shows a wide clear area, starting from the surface, around 155 nm thick, followed by a high contrast, irregular layer, identifiable by orifice-like dark features, which extend up to around 240 nm deep. Upon further inspection on the two HRTEM micrographs of the right, the clear layer seems to be amorphous-like, although it seems there are differences in crystallinity between atomic layers close to the surface (A) and close to the substrate (B). Judging by the FFT micrographs, the layers close to the surface seem to be fully amorphous, as there are no rings nor visible features in the FFT image. However, close to the substrate, around 155 nm deep, the FFT shows one main ring and several individual peaks, which would point out to the presence of randomly oriented nanocrystals. There seems to be a native oxide layer present on the surface, around 4 nm thick, hardly visible on HRTEM micrograph I of Fig. 4.5.

Using the same Ti dose, we increase the laser fluence up to 2.8 J/cm^2:

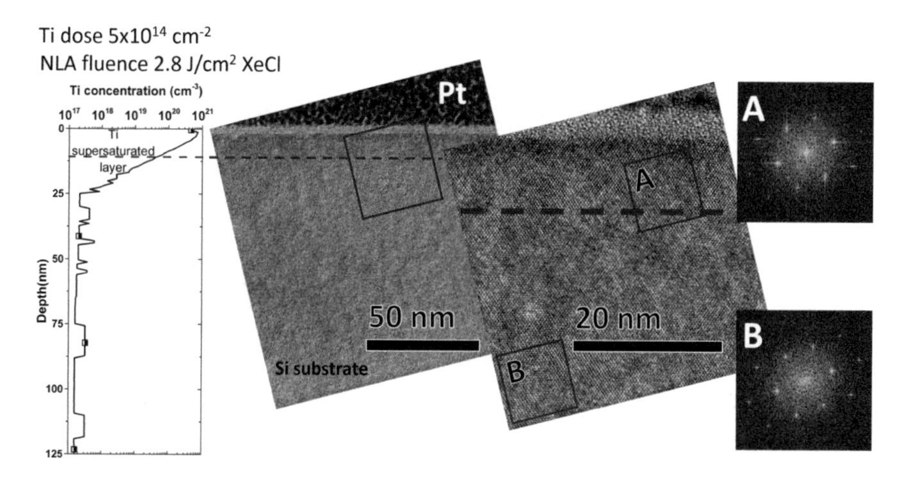

Fig. 4.4 From left to right: Ti profile distribution, overview TEM micrograph (at same scale as SIMS data), HRTEM micrograph of the red marked area on previous picture and two FFT of the areas marked as A and B of HRTEM micrograph. The horizontal red line indicates the limit of Ti supersaturating. Surface is on top of the pictures

There are differences with respect to the previously analysed TEM micrographs. The majority of the laser annealed area seems to be crystalline, as the FFT micrograph labelled as "B" shows with its discrete pattern. Visible in the TEM micrograph of the left there are several columnar structures, which have been identified as cellular breakdown structures, similar to the ones reported by Olea et al. with the short

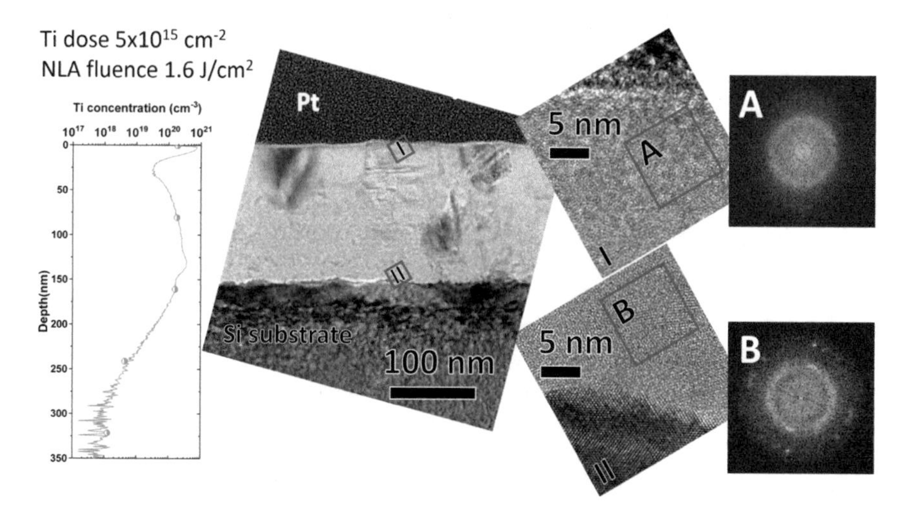

Fig. 4.5 From left to right: Ti profile distribution, overview TEM micrograph (at same scale as SIMS data), HRTEM micrograph of the red marked areas I and II on previous picture and two FFT of the areas marked as A and B of HRTEM micrographs

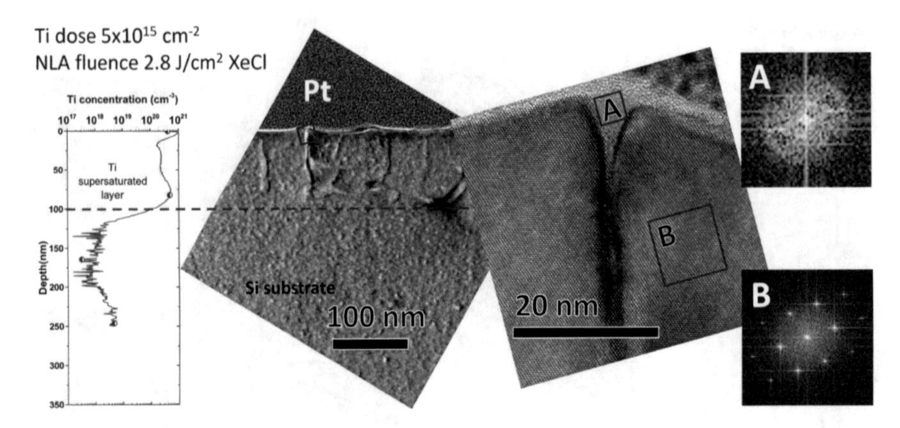

Fig. 4.6 From left to right: Ti profile distribution, overview TEM micrograph (at same scale as SIMS data), HRTEM micrograph of the red marked area on previous picture and two FFT of the areas marked as A and B of HRTEM micrograph

pulse duration laser and similar ion implantation parameters [3]. These structures average a length of around 85 ± 10 nm, shorter than the Ti supersaturated layer thickness, set to around 103 nm (Table 4.1). We show a close-up of one of those filamentary structures in the HRTEM micrograph on the right. Most of the area is monocrystalline, including most of the cellular breakdown structure, except on the top, where a trumpet-like structure opens up towards the surface. The contrast of the image in this area is clearly different to the rest, even different from the native oxide visible in the surface. With the aid of the FFT we see that this area is amorphous; only a featureless circle is observed. We rely on EDX measurements to identify the chemical composition of the trumpet-like structure and the surroundings.

We analyse the EDX spectra obtained in two different points of the sample. A first spectrum was obtained in an area close to the surface. The chemical composition is formed by silicon and oxygen atoms, in an approximate ratio of 83% and 17% respectively. Ti atoms were not detected in the examined area. The second spectrum was measured in the middle of the trumpet-like area, identified as amorphous by the FFT on the HRTEM micrograph on Fig. 4.6. This area is composed of 49% of Ti and 51% of Si, without any trace of oxygen.

The last sample with the middle dose is the sample with the highest laser fluence, 4.0 J/cm^2. The cross-section micrographs look similar to the sample with the same Ti dose but laser annealed at a lower laser fluence of 2.8 J/cm^2. In the sample annealed at 4.0 J/cm^2, the cellular breakdown structures are visible up to a depth of 100 ± 10 nm. The annealed layer seems more uniform, with less contrast than in the previous sample, which had a granular-like appearance. The crystal structure is monocrystalline on the whole annealed area. The crystal orientation along the CB filament is the same as the surrounding area. The EDX results inside the trumpet-like structure showed similar composition results: a 50% Ti 50% Si ratio, while oxygen atoms were detected in the surficial layer that covers the sample.

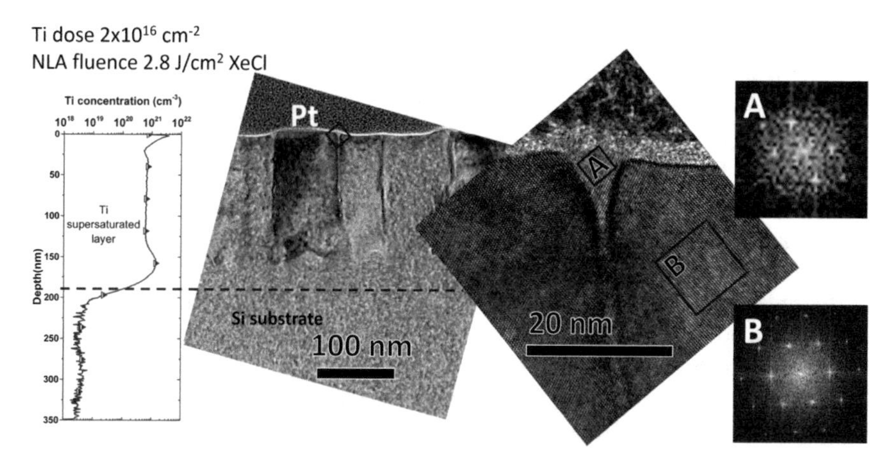

Fig. 4.7 From left to right: Ti profile distribution, overview TEM micrograph (at same scale as SIMS data), HRTEM micrograph of the red marked area on previous picture and two FFT of the areas marked as A and B of HRTEM micrograph. Surface is on top of the pictures, marked by the platinum layer used to protect the surface during FIB-SEM sample preparation. The sample is double Ti implanted with a total dose of 2×10^{16} cm^{-2} and a laser fluence of 2.8 J/cm^2

Finally, we analyse the sample with the highest Ti dose, 2×10^{16} cm^{-2}, laser annealed at 2.8 J/cm^2.

The TEM micrographs of Fig. 4.7 show similar results to what we observed on the samples with a Ti dose of 5×10^{15} cm^{-2}, laser annealed at 2.8 J/cm^2 and 4.0 J/cm^2: cellular breakdown structures, in this case of around 140 ± 15 nm long, and a monocrystalline structure in the whole implanted layer. In this case the trumpet-like structure atop of the CB filament seems mainly monocrystalline, except the upmost part, close to the native oxide, as deducted from the FFT of area labelled as "A" in Fig. 4.7.

Discussion on structural and compositional characterisation

SIMS results provided evidence that Ti was incorporated to the Si lattice in concentrations high enough to overcome the IB formation limit in all the samples studied in this section, with three different Ti dose, and three different laser fluences.

Before entering into more details with the laser annealed samples, it is necessary to make a paragraph explaining the melting and recrystallization processes on heavily implanted Si substrates, especially if this process produced an amorphous layer atop the substrate.

Amorphous silicon exhibits different optical and thermodynamic properties than its crystalline counterpart. According to several works, it is commonly accepted that amorphous silicon has a lower fusion point, around 225 ± 50 K lower than crystalline silicon [10], commonly accepted [1] to be 1683 K. Amorphous silicon has an absorption coefficient at $\lambda = 308$ nm similar to crystalline silicon, as it has been pointed out in other works [11, 12]. Therefore, the amorphous layer may melt first [13].

The melting and subsequent solidifying process of amorphous silicon atop a crystalline substrate is rather complex, as described in several works [14, 15]. When an amorphous layer has been melted, its solidification follows an "explosive recrystallization" process [15]; the solidification process starts shortly after the absorption of the laser light. Polycrystalline or crystalline phases are energetically more favourable [16], which means that polycrystals with varying grain size, depending on the melting conditions, should be the preferred regrown structures after the solidification process, even a fully monocrystalline layer. There are exceptions: if the solidification front speed is high enough (some references [17] set the limit at 15 m/s), the recrystallization process could result in an amorphously regrown layer, as the atoms would not have enough time to reorganise into more stable phases.

According to the literature, there are different minimum threshold laser fluences required to melt amorphous and crystalline silicon layers (due to their difference in the melting point), being the former considerably lower. Therefore, there would be a range of laser fluences in which it could be possible to melt the amorphous layer but not the crystalline substrate underneath. According to the cited references, there would be, simplifying, three different regimes, depending on the laser fluence: a first regime where the energy is not enough to melt the amorphous silicon (which we will refer in this text as the sub-melt regime), a second region where only amorphous silicon is melted (first crystallization), and a third one which would melt the amorphous silicon and, at least, part of the crystalline substrate (high energy). This is a first approach of the phenomenon, which will be further verified in Chapter 5.

Coming back to the "explosive recrystallization" process, it happens in both the first recrystallization and the high energy regimes. Other authors demonstrated [10] that the transition from amorphous to polycrystalline or crystalline phase is a transition of first order, which implies a latent heat of fusion that is released to the surroundings. This heat is transmitted mainly to the underlying atomic layers due to their higher thermal conductivity, when compared to the upper layers (air in this particular case). This heat has been found to be enough to melt again the remaining amorphous silicon, thus creating a melting propagation front that moves towards the substrate until the still-defective crystalline phase is encountered. What may happen later depends on the value of the laser fluence: if the laser fluence is not high enough to melt part of the crystalline substrate, the melting front would stop at the amorphous-crystalline (a–c) interface. The solidification front would start at this depth towards the surface, presumably as an amorphous or polycrystalline phase, as it could not take the substrate as a monocrystalline seed. If the laser fluence is high enough to reach the high energy region, the solidification process could take the monocrystalline structure from the partially melted substrate, possibly recrystallizing as a monocrystalline structure.

With these ideas in mind it is possible to further interpret the experimental results. The analysis of the samples annealed using a NLA process at different laser fluences showed that there is a redistribution of Ti atoms inside the Si lattice, possibly due to the melting process. Ti diffusion coefficient in liquid Si has been proven to be very high [18]. In other words, Ti atoms are extremely mobile in the liquid (molten) Si phase, while being comparatively slow on the solid phase. Thus, Ti atoms would tend to stay in the molten phase as long as possible, moving towards the surface, following the solidification front, which starts from the melting depth. This redistribution effect, already mentioned as the snow-plow effect, is clearly visible in the analysed samples. Thus, the point at which both the laser annealed Ti profile and the as-implanted profile overlap could be used as a good estimation of the melting depth of the laser. A similar approximation has been already reported with other transition metals ion implanted and subsequently laser annealed on Si, as copper [19]. Those melted depth estimation values are displayed in the fourth column of Table 4.1, and there are two trends observed, as a function of the Ti dose and as a function of the laser fluence:

- Increasing the Ti dose (at a fixed laser fluence) increases the melted depth.
- Increasing the laser fluence (at a fixed Ti dose) increases the melted depth.

The first affirmation is related to the influence of the thickness of the amorphous layer in the thermodynamics of the melting and solidification process. Higher Ti doses lead to thicker amorphised layers. If the a–c interface is deeper, it would drive deeper the melting front.

The influence of the laser fluence on the melting depth is also consistent with the explosive recrystallization theory. TEM micrographs of the sample annealed at 1.6 J/cm^2, in Fig. 4.5, showed a damaged layer, around 155 nm thick, where we found different levels of crystallinity. The presence of nanocrystals possibly indicates that there may have been a recrystallization process at least up to their depth. This fact would point out to a laser fluence too low to reach the melting threshold of the crystalline phase, but high enough to melt the amorphous layer. The results obtained by SIMS, where we found a redistribution of Ti atoms from their as-implanted positions, also point out to a laser process that was able to melt only the amorphous layer (what has been described before as the "first recrystallization" regime). The melting process displaced Ti atoms up to around 140 nm deep, which is observable on the SIMS profile of the annealed sample at 1.6 J/cm^2. This estimated melting depth is close to the depth of the a–c interface, backing the hypothesis of the explosive recrystallization process. The difference in crystallinity between the deeper (nanocrystals) and the shallower (amorphous) part of the damaged layer could be explained by an increase in the solidification speed when approaching the surface [20], combined with the fact there was not a monocrystalline seed to regrow from.

The sample annealed at 2.8 J/cm^2 exhibited a monocrystalline structure, which points out to a laser melting process in the high energy region. However, according to SIMS data, the melting depth was not high enough to fully melt the Ti implanted layer, which may result in End Of Range (EOR) defects, damage caused by the ion implantation that has not been correctly annealed by the laser [21], which were visible on cross-section TEM micrographs at around the same depth (Fig. 4.6) as black dots.

The analysis of the compositional and structural properties led us to estimate the minimum laser fluence needed to produce a laser annealed monocrystalline layer. Knowing the threshold of monocrystalline recrystallization is of utmost importance to design devices based on Ti supersaturated Si. It is possible to estimate this threshold using Raman spectroscopy. We observed that samples annealed at 2.2 J/cm^2 displayed a Raman shift curve half-way between the as-implanted sample and the crystalline reference. The next higher laser fluence was 2.8 J/cm^2, which was monocrystalline, according to cross-section HRTEM. Therefore, we conclude that the threshold of monocrystalline crystallisation should lie between 2.2–2.8 J/cm^2 for the three Ti doses studied. Thus, laser fluences equal or lower than 2.2 J/cm^2 would lie in the sub-melt or first crystallisation regime (together they conform the "low energy" regime), while values equal or higher than 2.8 J/cm^2 would lie in the high energy regime. Raman spectroscopy measurements and the micrographs obtained by cross-section HRTEM are consistent. The sample implanted at a dose of 5×10^{15} cm^{-2}, laser annealed with the lowest fluence exhibited a smaller peak at around 515 cm^{-2}, which could be related to the presence of amorphous or polycrystalline silicon [22], which we have verified using TEM.

We used TEM to obtain more information on CBD structures. The sample with the lower Ti dose did not exhibit any trace of cellular breakdown, yet SIMS showed a profile with enough Ti concentration to achieve supersaturation. For higher Ti doses and using EDX we estimated the composition of the trumpet-like structure, which was composed around 50% Ti and 50% Si. The EDX measurements only show a significant amount of Ti accumulation in the trumpet-like structure, which is at most 10 nm long, compared to the Ti Implanted Layer (TIL) thickness, in the order of 100–120 nm, as obtained from SIMS. The contribution of the Ti atoms from the trumpet-like structures could be important only in the first 10 nm, but could not explain the high Ti concentrations at points deeper than the trumpet-like structure.

We aim to obtain thick Ti implanted layers with good crystal quality to fabricate photodiodes with increased absorption, while keeping the noise and dark current at low levels. So far, we have measured that the TIL has reached Ti concentrations higher than the IB formation limit in layers between 10 and 195 nm thick, which makes the material suitable to absorb sub-bandgap photons. The crystal quality of samples annealed with laser fluences higher than 2.8 J/cm^2, what we have labelled as the high energy regime, showed monocrystalline regrown layers in the three Ti doses under study. Therefore, the Ti supersaturated material seems suitable to fabricate photodiodes from the compositional and structural point of view.

4.3.2 Inhomogeneities on the Ti Supersaturated Material

As with the KrF laser, we perform the analysis of the different inhomogeneity sources: craters and laser overlapping lines. Cellular breakdown structures have been previously described. Given the previous inputs coming from the inhomogeneity study described in Sect. 3.4, we increased the cleanliness during the sample preparation. Before packaging the samples and sending them for NLA processing, we cleaned the samples using acetone and IPA, drying the samples after with a nitrogen blow, which helped to remove the possible solid deposited particles on the wafer surface. The aim of this extra cleanliness was to test if the density of craters could be lowered,

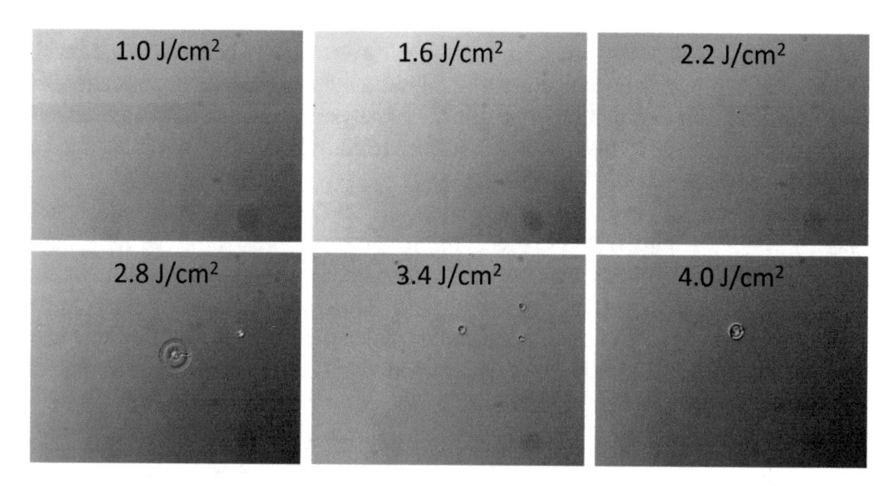

Fig. 4.8 Craters and NLA overlapping lines (visible in the first picture), as seen from an optical microscope, of six samples double Ti implanted with a total dose of 2×10^{16} cm^{-2}, laser annealed with the XeCl laser at different laser fluences

as we suspect the craters may be formed due to solid particles present in the surface at the moment of the NLA process.

Figure 4.8 shows the surface of six samples, with a Ti total dose of 2×10^{16} cm^{-2}, laser annealed with the XeCl 155 ns laser. Comparing with Fig. 3.5, where we took the same magnification on the optical microscope, the crater density has decreased considerably on the set of samples annealed with the XeCl laser. The density of the craters is greatly reduced, even though the Ti dose on these samples is higher than in the samples shown in Fig. 3.5. In fact, in this set of samples, we could not see any correlation between the density and average size of the craters and the laser fluence nor the Ti concentration, as opposed to what we observed when using the KrF laser.

With respect to NLA overlapping lines, since the exposed area was 10×10 mm, the quantity of lines is greatly reduced. They are barely visible only in the upper left photograph of Fig. 4.8.

Therefore, we conclude that the XeCl laser, possibly helped by a better cleaning process, may introduce less sources of inhomogeneities in the material, a desirable effect to fabricate a pixel matrix, where all pixels are expected to behave similarly.

4.3.3 *Optical Properties*

The analysis contained in this section is based on Transmittance-Reflectance (T-R) measurements, as described in Sect. 2.4.1. Those measurements will provide information about the absorption processes of Ti implanted samples, which may determine if there is sub-bandgap absorption, necessary to fabricate a photodiode responsive in the NIR.

Each quarter of wafer was measured by pointing the incident beam from the Perkin Elmer 1050 instrument on the centre of the laser annealed area (beam diameter less than 5 mm diameter over a 4 cm^2 area). First, we analyse the transmittance, reflectance and absortance curves of three samples with increasing Ti dose but laser annealed at the same laser fluence of 2.8 J/cm^2 (Fig. 4.9).

The transmittance curves show that increasing the Ti dose decreases the transmittance in the sub-bandgap region. The Ti implanted samples show higher reflectance in the visible part of the spectra than the reference. In the same range we observe oscillations, except for the case of the lower Ti dose, where a similar curve to the Si reference is found. The absortance curves, calculated as A = 1-T-R, show sub-bandgap absorption, which increases along with the Ti dose. The sample implanted

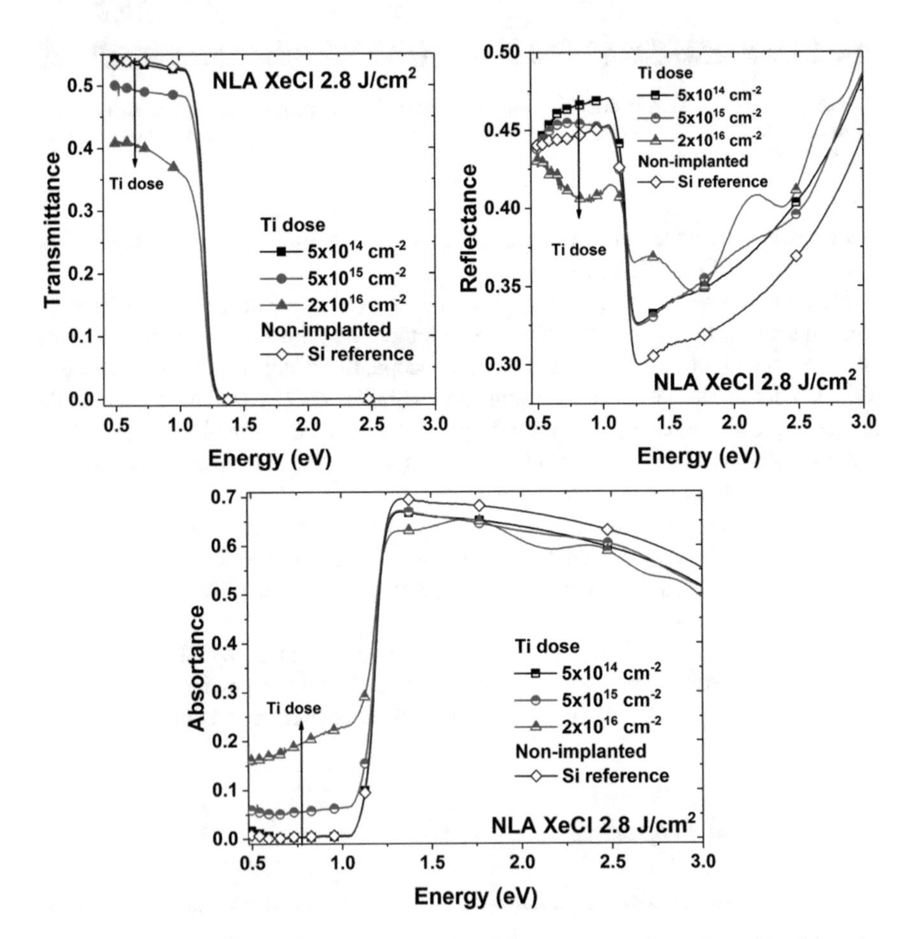

Fig. 4.9 Transmittance (upper left), reflectance (upper right) and absortance (bottom) as a function of the photon energy for NLA processed samples at a fluence of 2.8 J/cm^2, for Ti implanted samples at three different Ti doses: 5×10^{14}, 5×10^{15} and 2×10^{16} cm^{-2}

with the lower Ti dose shows low absortance in the sub-bandgap region, going below the noise level, set to 8×10^{-3} of absortance, as it happens with the reference sample.

After, we study the effect of the laser fluence in the optical properties of samples implanted at the middle Ti dose, 5×10^{15} cm^{-2}, annealed using three different laser fluences of 1.6, 2.8 and 4.0 J/cm^2 (Fig. 4.10).

The transmittance curves show little differences in the sub-bandgap region; the laser fluence barely affects the transmittance. The reflectance curves show the same average reflectance values for Ti implanted samples in the visible part of the spectra. However, the oscillations strongly depend on the laser fluence: the amplitude of the oscillation decreases with increasing laser fluence, until they almost disappear in the case of 4.0 J/cm^2. The reflectance in the sub-bandgap region shows an increase in

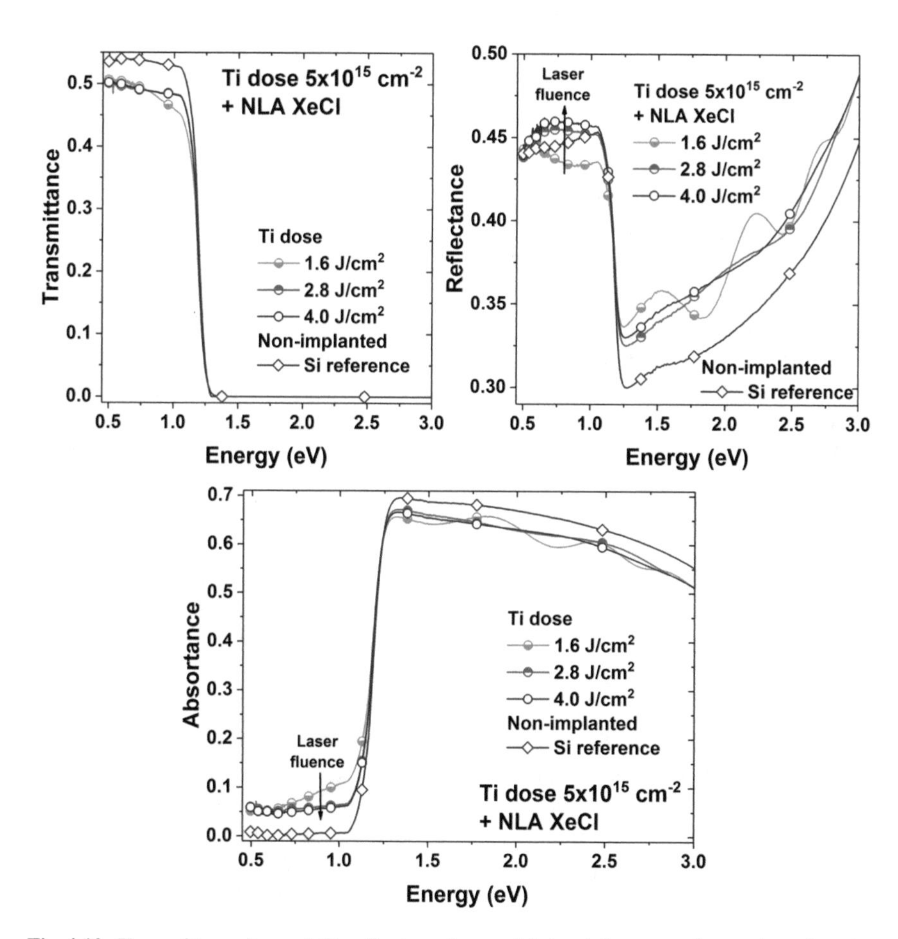

Fig. 4.10 Transmittance (upper left), reflectance (upper right) and absortance (bottom) as a function of the photon energy for NLA processed samples at different fluences, for Ti implanted samples at a fixed Ti dose of 5×10^{15} cm^{-2}

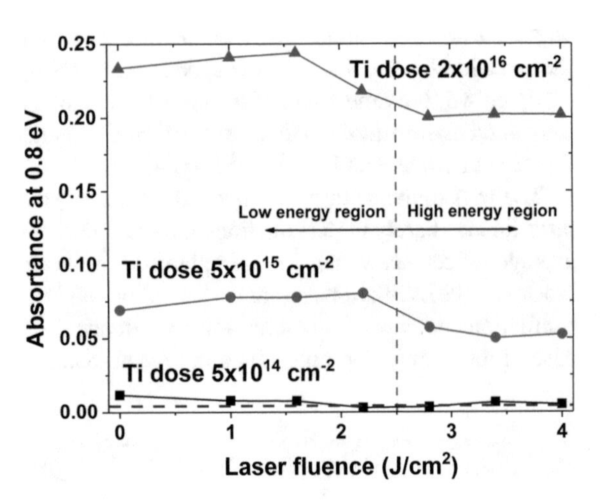

Fig. 4.11 Absortance at a photon energy of 0.8 eV for Ti implanted samples at three different doses, as a function of the laser fluence. The violet dashed line indicates the absortance measured in the non-implanted, non-annealed Si sample as a reference

the reflectance along with the laser fluence. In the case of the absortance, we observe sub-bandgap photon absorption in all the Ti implanted samples.

After showing the absortance curves as a function of the photon energy, we analyse the effect of both the Ti dose and the laser fluence in the absortance at the key photon energy of 0.8 eV (1.55 μm), widely used in the SWIR for fibre optic telecommunications (Fig. 4.11).

The absortance values at a photon energy of 0.8 eV seem to decrease slightly for higher laser fluences, as compared to the as-implanted values. The low energy and high energy regimes are distinguishable in this graph; low laser fluences ($F \leq 2.2$ J/cm^2) show higher absortance values than those sample within the high energy regime ($F \geq 2.8$ J/cm^2).

Absortance is a first approach to obtain an estimation of the absorption properties of a sample. The absorption coefficient is related to the imaginary part of the complex refractive index of the material. It is possible to estimate both parts of the complex refractive index using the T-R measurements, and applying a four-layer model described by Maley et al. [23]. The four-layer model describes two scenarios at each interface between layers: the coherent limit, which considers multiple reflections between two layers of different complex refractive indices N_i, including possible interferences, and the incoherent limit, in which the reflections at the interface are negligible. The application of the former requires, at least, a difference in the real part of the refractive index, while the latter assumes both real parts are equal. The use of either approximation may depend not only on the possible difference of refractive indices, but also on the morphology of the interface between layers. If the interface is rough, the incoming rays would reflect in random directions, leading to an absence of coherence between reflected rays, halting the observation of interference patterns in reflectance curves, even if both the imaginary and the real part of the refractive index are different at either side of the interface. Despite the higher complexity of the coherent limit as compared to the incoherent limit, its main advantage over the latter

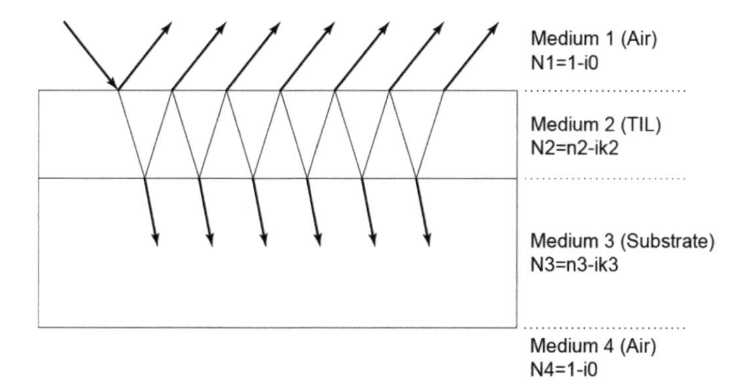

Fig. 4.12 Four layer model schematic in the coherent limit. Light rays are indicated in non-normal angle for clarity

is that it separates the contribution of the thickness and the absorption coefficient, estimating more precisely each parameter.

In the following, we will consider the coherent limit, decision motivated by the apparition of interference patterns in the reflectance curves of the Ti implanted samples annealed with low laser fluences. In our particular application of the four-layer model, the first layer would be air, where a light beam is directed in normal incidence to a second absorbing layer (the Ti implanted layer in our case), located atop a third layer (the Si substrate). The last layer would be air again.

The model uses complex numbers to perform the mathematical treatment of the data; the complex refractive index is indicated as N_i (see definition in Fig. 4.12). In Maley et al. [23], they developed a model in which the substrate was non-absorbent in the IR region, as they studied the infrared transmission of amorphous layers atop transparent substrates. Here, we extend the model to the visible range, where we cannot assume the substrate as non-absorbent in the whole range. The equations used to fit the model can be found on Annec C.

We use as inputs of the model the complex refractive index and the thickness of each layer. We implemented a numerical solution to the equations using the complex refractive indices of each layer. In the numerical model, we define the error of the transmittance and the reflectance as:

$$\Delta T = \left(T_{exp} - T_{model}\right)^2 \tag{4.1}$$

$$\Delta R = \left(R_{exp} - R_{model}\right)^2 \tag{4.2}$$

$$E_{total} = \Delta T + \Delta R \tag{4.3}$$

Where T_{exp}, R_{exp} are the experimental values, and R_{model}, T_{model} are defined in Annex C. The numerical model is set to minimise the total error, E_{total} by modifying the complex refractive index of the implanted layer, along with its thickness. For the silicon substrate, we use the tabulated values of the complex refractive index in the region of interest (from 300 to 2500 nm of wavelength) from Green et al. [11] and Edwards and Ochoa [24]. To model the complex refractive index of the TIL (N_2 in the equations of the model), we assumed a Cauchy dispersion for the real part n_2, and a three-order polynomial dependence for the imaginary part, as seen in other works [25]. More details in Annex C.

We applied the same model to the 21 Ti implanted samples. From the fitting to the model we obtain the real and imaginary parts of the refractive index (and hence the absorption coefficient), along with the thickness of the TIL. As in the rest of this section, only a few representative curves will be shown (Fig. 4.13).

As restrictions, we impose that neither d, n or k can be negative at any given wavelength. Using the four-layer model we could fit the 42 (two per each sample) curves with relative low error values. The quality of the fitting varied between samples,

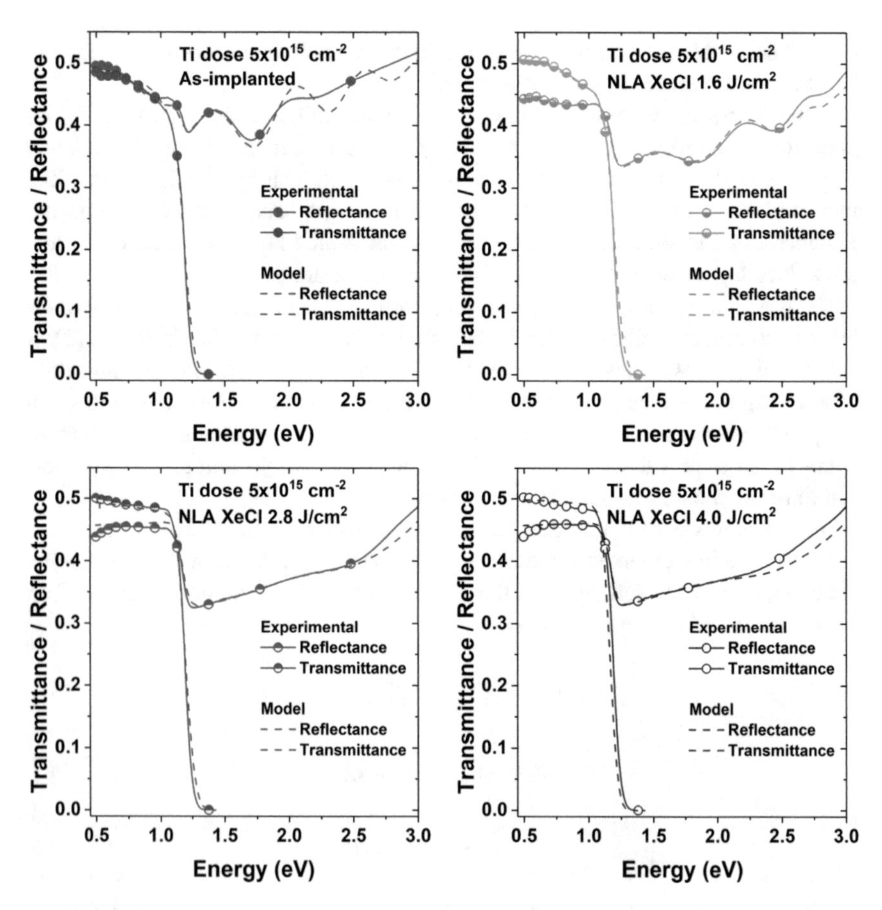

Fig. 4.13 Four-layer model applied to four samples double Ti implanted at a total dose of 5×10^{15} cm^{-2}, as-implanted (upper left) and laser annealed at 1.6 (upper right), 2.8 (lower left) and 4.0 J/cm^2 (lower right)

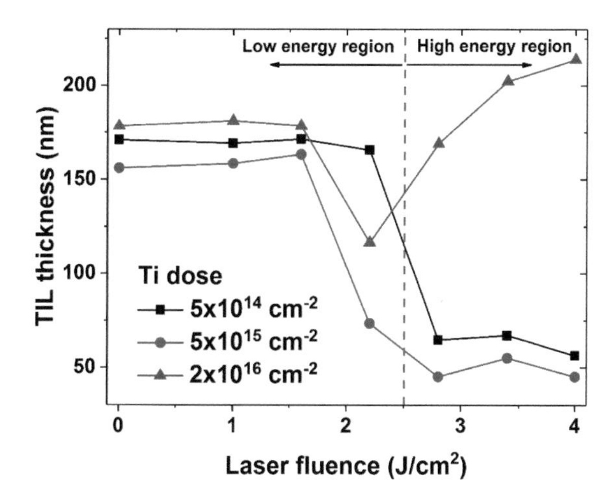

Fig. 4.14 Optically absorbing Ti implanted layer thickness as a function of the laser fluence for three different Ti doses. The as-implanted samples are shown at 0.0 J/cm^2

but on average the model (applying the coherent limit) simulated better the curves of the as-implanted and the low laser-annealed samples. At high laser fluences, the oscillations seen in reflectance drastically reduce their amplitude, making it more difficult to obtain a good fitting.

First, we study the influence of the Ti dose and the laser fluence on the TIL thickness.

In Fig. 4.14, the as-implanted and low laser annealed samples exhibit a rather constant thickness value, between 155 and 185 nm, depending on the Ti dose, up to a certain value around 2.2 J/cm^2 of laser fluence, which is the limit between the low energy and high energy regimes observed in the structural characterisation. For samples with the low and middle Ti doses, laser annealed at fluences equal or higher than 2.8 J/cm^2, the thickness drops down to around 40–70 nm, the opposite of what happens in samples implanted with the higher Ti dose.

All the samples shown in Fig. 4.15 show sub-bandgap absorption coefficients higher than the reference, in the order of 10^2–10^4 cm^{-1}. In the left graph, we observe how the increase in the Ti dose leads to an increase of the absorption coefficient in the sub-bandgap region and part of the energies over the bandgap. It is interesting to note that as the Ti dose increases the absorption coefficient increases in a wider photon energy range. For example, in the case of the lower dose, the absorption coefficient is higher than the Si reference from 0.5 up to 1.5 eV, while in the case of the highest dose the absorption coefficient of the implanted sample overcomes the one of bare silicon from 0.5 eV up to 2.9 eV, which is well within the visible range.

The absorption coefficient seems less affected by the laser fluence than by the Ti dose. The sample annealed with the lower laser fluence, 1.6 J/cm^2, shows higher curvature than the other samples, showing a local minimum around 2.5 eV, which is caused by the fitting to a third-degree polynomial. The samples annealed with higher laser fluences display similar absorption coefficients, although the higher values in the sub-bandgap region are reported by the sample annealed at 2.8 J/cm^2. Finally, we represent the real part of the refractive index as a function of the energy.

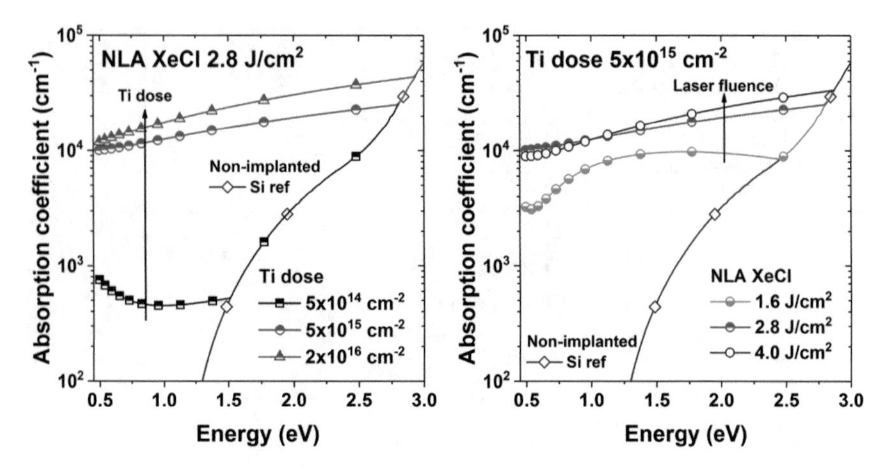

Fig. 4.15 Absorption coefficient in Y-log scale, as a function of the photon energy for three samples implanted at different Ti doses, laser annealed at 2.8 J/cm^2 (left) and three samples implanted at a fixed Ti dose of 5×10^{15} cm^{-2}, annealed at three different laser fluences. A non-implanted, non-annealed Si reference sample is included for comparison

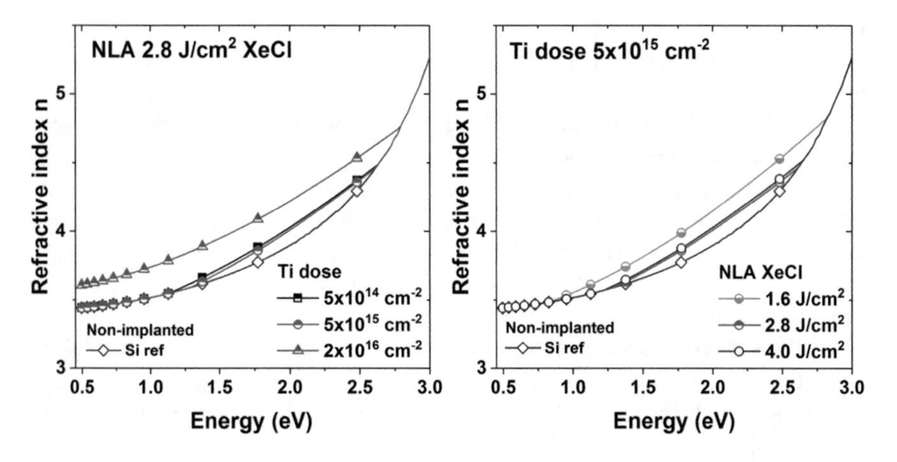

Fig. 4.16 Real part of the refractive index as a function of the photon energy for Ti implanted samples at three different Ti doses, annealed at a fixed laser fluence of 2.8 J/cm^2 (left), and three samples Ti implanted samples with a total dose of 5×10^{15} cm^{-2}, laser annealed at different laser fluences (right). The reference Si sample is shown for comparison in both graphs

All the implanted samples, using the same model, showed higher refractive index than the reference. On average, the samples with the highest Ti dose had higher refractive index values than the rest of the samples, although the difference between the lower and the middle dose were not as clear. In the case of the effect of the laser fluence, increasing the laser fluence may lead to lower refractive indices.

Discussion

The optical properties of the Ti implanted samples are considerably different from the reference silicon: the transmittance of the Ti implanted samples was always lower than that of the reference. Besides, the reflectance in the sub-bandgap region was lower for most Ti implanted samples, while showing higher reflectance values in the visible range. Reflectance measurements also displayed an unexpected behaviour (from our previous knowledge, see Olea et al. [3]) in as-implanted and laser annealed samples with low laser fluences: oscillations caused by light interferences, possibly due to the interface of the TIL and the substrate. The interpretation of these measurements led to two main observations:

- Interferences are likely related to a difference in the real part of the refractive index between the TIL and the substrate.
- The imaginary part of the TIL must be higher than that of the reference in the sub-bandgap region, due to the higher absortance measured.

The examination of the absortance measurements led to the conclusion that higher Ti doses lead to higher absortance values in all the studied cases. With respect to the influence of the laser energy density, we observed that, in general, lower fluences lead to higher absortance values.

Considering the information already described in the compositional and structural characterisation section, it is possible to relate the absortance values to the Ti concentration present in the TIL. In SIMS measurements, we observed that higher Ti dose led to higher Ti volumetric concentrations. We also measured that increasing the laser fluence decreased the retained Ti dose, in a linear way (see last column of Table 4.1). We observed that the absortance decreased when increasing the laser fluence, in line with our previous reasoning. In particular, for the two higher Ti doses, we found two regions with rather constant absortance values, with the transition located at 2.2 J/cm^2. Laser fluences lower than this value offered higher absortance values than the samples annealed at higher fluences. Taking into account the TEM micrographs and the Raman spectra measured on the Ti implanted samples, we suspect this double behaviour could be caused by two differentiated absorption processes: at low laser fluences (which we have labelled as the sub-melt and first crystallisation regimes before. This regime also includes as-implanted samples), the absorption processes could be driven by non-annealed defects coming from the ion implantation process. For example, the sample Ti implanted with a total dose of 5×10^{15} cm^{-2}, laser annealed at 1.6 J/cm^2 did not show a monocrystalline structure, but polycrystalline or amorphous. As stated in other references, the absorption coefficient of amorphous silicon is drastically different than that of monocrystalline silicon [12], which could explain why the absortance and also the absorption coefficient were different for this sample, when compared to samples annealed at higher laser fluences (around 2.8 J/cm^2 or higher, within the high energy regime). The second absorption mechanism would be the presence of Ti atoms. At higher laser fluences, almost all damage coming from the ion implantation has been annealed (judging from TEM and Raman data), revealing the contribution of Ti atoms to the optical absorption processes.

We continued the analysis of the T-R measurements by applying a four-layer model to the experimental data, where we assumed a coherent reflection of incoming light beams, at the interface between the TIL and the substrate. The application of the four-layer model in the coherent limit allowed us to estimate the real and imaginary part of the refractive index, as a function of the photon energy, for each Ti implanted sample. The model uses the thickness of each layer in its calculations: the thickness of the substrate is known, and the thickness of the Ti implanted layer is a parameter used to fit the model. In the following, we will consider the fitted parameter as the thickness of the "optically active" layer, not necessarily related to any given value of Ti concentration.

The model accurately reproduced most of the reflectance peaks, in the as-implanted samples and in those annealed at laser fluences lower than 2.8 J/cm^2 for all Ti doses. At higher laser fluences, the samples with the highest Ti dose still showed oscillations in the reflectance curves, while the middle and lower dose showed damped oscillations. In the latter group of samples, the four-layer model in the coherent approximation showed stronger limitations, as the decrease in the amplitude of the oscillations made it difficult to obtain a more accurate fitting to the model. This is clearly visible in the thickness of the optically active layer (Fig. 4.14): it does not follow the expected behaviour: the lowest Ti dose should lead, according to SIMS measurements, to thinner Ti implanted layers. The lower dose exhibits thicker optically active layers than the middle dose, which does not agree with SIMS measurements; the implementation of the model may require further tuning to increase its accuracy. The optically active TIL thickness values (see Fig. 4.14) point out again to two different regions: the low energy regime, for laser fluences equal or lower than around 2.2 J/cm^2, and the high energy regime, for values higher than 2.2 J/cm^2.

If we go back to the compositional and structural characterisation (Sect. 4.3.1), we observe that oscillations are appearing in what we called the sub-melt and first recrystallization regimes, where we found poor crystal quality (amorphous or nanocrystalline structures). The amplitude of the oscillations decreases when the laser fluence is increased to values equal or higher than 2.8 J/cm^2, which coincides when the TIL has recrystallized as a monocrystal. Therefore, the oscillations could originate from the interface between the poorly recrystallized layer (laser fluences in the low energy regime, including the as-implanted sample) and the crystalline substrate, which physically corresponds to the a–c interface, located between 130–200 nm for the different Ti doses (see for example, TEM micrograph in Fig. 4.5). Thus, when the TIL is monocrystalline, there would be no abrupt interface, losing the coherence between reflected rays, leading to lower thicknesses. Lowering the thickness is the mathematical solution used in the model to decrease the number of interference peaks (see Fig. 4.14). This would explain why the model showed limitations when applied to samples submitted to high laser fluences, which showed damped oscillations. It is well known that the interference phenomena are maximum when the thickness of a layer is in the order of $\lambda/4$ of the illumination. That would lead to wavelength values in the order of 400–800 nm, within the visible range, where oscillations were experimentally observed. Therefore, the reflectance measurements could be used to estimate the crystal quality of Ti implanted layers after a laser process in a fast and non-destructive way, provided that the energy of implantation is high enough to produce amorphous layers sufficiently thick. The different case of the highest Ti dose, where we observed an increase in the thickness of the optically active layer could be explained by a much higher real part of the refractive index, even after recrystallizing as a monocrystal (see Fig. 4.16 left).

The fitting of the model to the experimental results led to the calculation of n and k as a function of the wavelength. The real part of the refractive index was found to be dependent on both the Ti dose and the NLA process. All the Ti implanted samples showed higher refractive indices than the reference in most of the studied range. If we set the Ti dose at 5×10^{15} cm^{-2} and observe the influence of the NLA process in the refractive index, we see it decreases when increasing the laser fluence.

The absorption coefficient is strongly influenced by the Ti dose. In Fig. 4.15 left we see a clear dependence on the Ti dose and the absorption coefficient in the sub-bandgap region: the higher the dose, the higher the absorption coefficient, similarly to what we observed when analysing the absortance curves. Although the absortance curves showed that the sample with the lower Ti dose did not absorb sub-bandgap photons over the noise level, the four-layer model led to absorption coefficients higher than the reference in order to explain the experimental curves. This is consistent with the low absorption coefficient of these samples (in the order of 400 cm^{-1}) and their low TIL thickness, in the order of 10 nm, according to SIMS (Fig. 4.2). The effect of the laser fluence is also noticeable (Fig. 4.15 right): the sample annealed with the lower fluence, 1.6 J/cm^2, show a different curvature and lower absorption coefficient. The samples annealed at 2.8 and 4.0 J/cm^2 showed similar values, which again points out to a stabilisation of the optical properties at high laser fluences (as it happened with the refractive index) in the high energy regime. In this regime, the absorption coefficient showed a rather flat curve, which overlapped with the reference sample at different energies. According to the fitting of the experimental data to the four-layer model, the absorption coefficient of Ti implanted samples could be increased not only in the sub-bandgap region, but also at photon energies over the bandgap. For example, the sample implanted with the highest Ti dose, annealed at a laser fluence of 2.8 J/cm^2 (Fig. 4.15 left) overlaps with the reference Si sample at 2.95 eV ($\lambda =$ 420 nm), well inside the visible range. This result is a novelty in our research group, as the previous measurements were performed in a narrower energy interval.

Summing up the discussion on this section, we found sub-bandgap absorption on all Ti implanted samples, which is the first step to detect sub-bandgap photons. The doses of 5×10^{15} and 2×10^{16} cm^{-2} exhibited higher absorption coefficients, in the order of 10^3–10^4 cm^{-1} at 0.8 eV, which added to the fact of their increased Ti supersaturated thickness, makes them more suitable for IR photodetectors than the lower Ti dose. The application of the four-layer model allowed us to estimate, based solely on the presence of interferences in the reflectance curves, the minimum threshold to produce a monocrystalline layer in samples with Ti doses equal or lower than 5×10^{15} cm^{-2}.

4.3.4 *Electrical Properties: Four-Point Measurements*

The electrical properties of Ti implanted samples have been studied in the past in our research group, profusely described in the theses of J. Olea [26] and E. García [27] and in several published works [28–32], where a bilayer model was developed. A brief summary of the bilayer model can be found in Annex D. In this sub-section, we detail the electrical measurements obtained by four-probe measurements, i.e. sheet conductance, sheet concentration and Hall mobility values using the van der Pauw configuration, as previously described in Sect. 2.4.4, on n-type and p-type substrates. This characterisation will be referred to as "horizontal" characterisation in the following.

4.3.4.1 N-Type Si Substrates

First, we will study the samples based on the n-type Si substrates, code N1 (Table 2.1), laser annealed with the XeCl laser. The aim is to check if the behaviour of these samples is similar to what our research group observed with the KrF laser. We analysed seven samples: the reference and two samples laser annealed at 2.8 and 4.0 J/cm^2 for each Ti dose. We represent the sheet conductance, the Hall mobility and the sheet concentration, for the three Ti doses used in this section.

Figure 4.17 left shows the sheet conductance as a function of the temperature. Three different regimes can be observed: at high temperatures (above 250 K for all samples) the three curves go parallel, almost overlapping with the reference. The

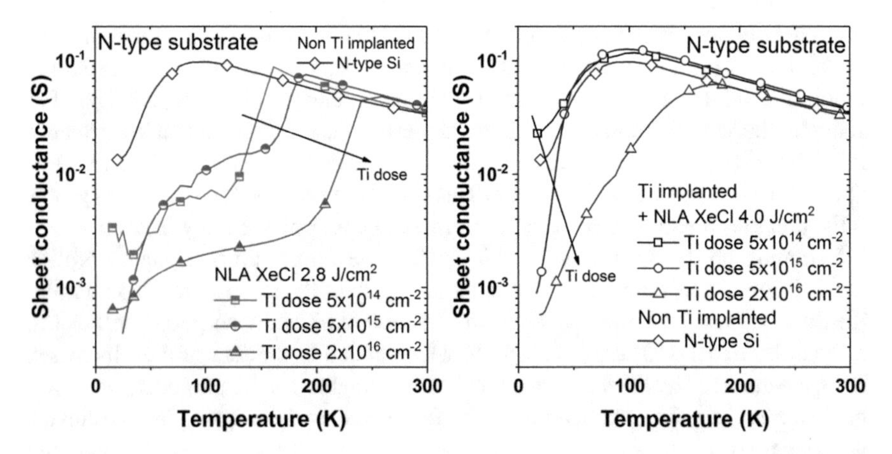

Fig. 4.17 Sheet conductance as a function of the temperature for Ti implanted samples at three different doses and a non-Ti-implanted reference. Ti implanted samples shown in the left figure were laser annealed using 2.8 J/cm^2, while the ones represented on the right were laser annealed using 4.0 J/cm^2

second region is found at low temperatures (below 100 K for the three samples), where low sheet conductance values are found, lower than the reference. The third region is located between the two previous regions, where there is a transition from high to low sheet conductance. Each Ti dose shows a different transition temperature, being higher as the Ti dose increases (see arrow in Fig. 4.17 left). Special attention must be put when examining the curve having a Ti dose of 5×10^{14} cm^{-2} as it is particularly noisy for temperatures lower than 140 K.

Figure 4.17 right shows the conductance as a function of the temperature but for samples annealed at 4.0 J/cm^2. The conductance behaves differently in this second set of samples, as compared to the left figure. The lower and middle dose go almost parallel to the reference in the whole temperature range, except for temperatures lower than around 40 K, where the Ti implanted samples exhibit higher sheet conductance values except in the case of the lower Ti dose. The highest dose shows a transition at temperatures around 175 K. In the very low temperature region, the sheet conductance of the Ti implanted samples decreases as the Ti dose is increased.

Figure 4.18 shows the absolute value of the Hall mobility as a function of the temperature. The left graph, as in the previous figure, shows the set of samples laser annealed at 2.8 J/cm^2. The mobility curves for the Ti implanted samples annealed at 2.8 J/cm^2 show two different transition temperatures, leading to up to five different regimes. At temperatures below 60 K, the Hall mobility is very low, compared to the rest of the temperature range. This region points out to lower mobility values as the Ti dose increases, where some samples exhibited a change in the sign of the Hall voltage with respect to the n-type substrate, being their conduction process dominated by holes. Then, the three samples show a first transition zone, where the Hall mobility increases. The lowest Ti dose seem to show the same transition zone,

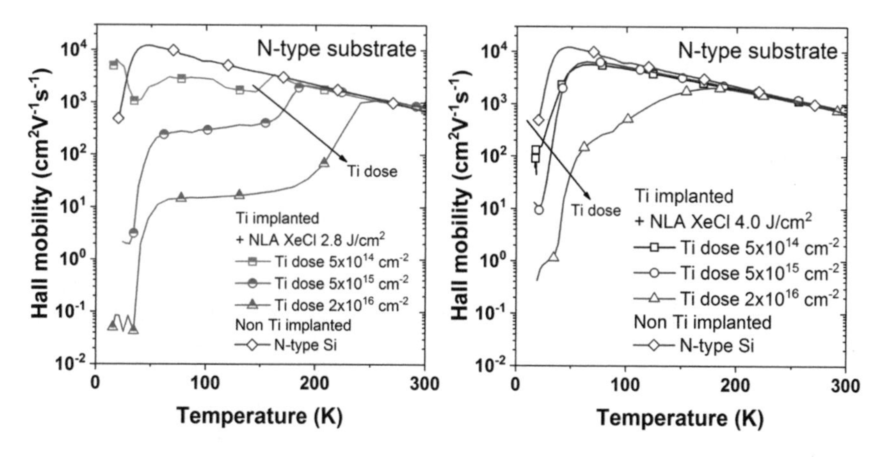

Fig. 4.18 Absolute value of Hall mobility as a function of the temperature for Ti implanted samples at three different doses and a non-Ti-implanted reference. Ti implanted samples shown in the left figure were laser annealed using 2.8 J/cm^2, while the ones represented on the right were laser annealed using 4.0 J/cm^2

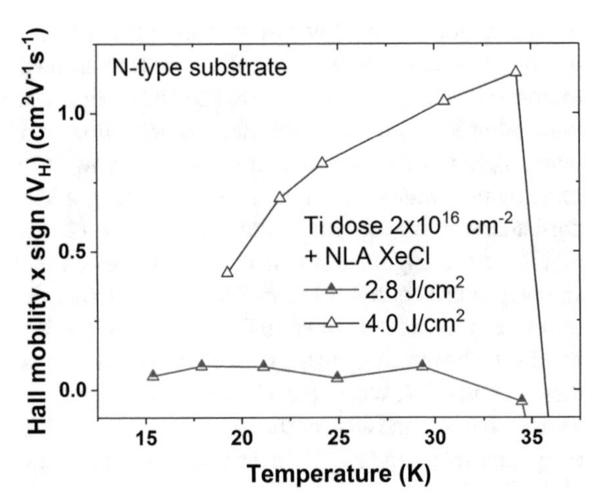

Fig. 4.19 Product of the Hall mobility and the sign of the Hall voltage as a function of the temperature for samples Ti implanted at a dose of 2×10^{16} cm^{-2}, subsequently NLA at different laser fluences. Substrate is n-type. At low temperatures, positive values of the Hall voltage were obtained

although the curve of this sample is particularly noisy and must be examined with caution for temperatures lower than 140 K. The three doses exhibit the same Hall voltage sign as the substrate, indicating that the conduction is dominated by electrons. After, we find an intermediate zone, where the mobility is almost constant. The Hall mobility in this region is strongly dependent on the Ti dose. The second transition occurs between 150 and 250 K, being the transition temperature higher as the Ti dose increases, as it was observed in the sheet conductance curves. At high temperatures, higher than 250 K for all the samples, the mobility values are close and go parallel to the reference substrate.

Figure 4.18 right shows a set of three Ti implanted samples laser annealed at 4.0 J/cm^2. The two samples with the lower and the middle dose exhibit curves similar to the reference, differing only at low temperatures, where they show considerably lower Hall mobility values. The sample with the higher dose showed a transition zone similar to what was observed on Fig. 4.17 right. In this low temperature region (below around 50 K) the mobility is dependent on the Ti dose: higher Ti doses lead to lower mobility values. It is worth studying the product of the mobility and the Hall sign at the regime of low temperatures, as some of the samples showed a change in the Hall sign for temperatures lower than 50 K (Fig. 4.19).

From the six samples measured, only two showed a change in the Hall voltage at low temperatures: the two samples having the highest Ti dose, which also showed the transition regime more clearly in Fig. 4.18. The positive mobility points out to a conduction mechanism dominated by holes, as opposed to the conduction dominated by electrons of the n-type substrate, in line to what was observed previously in our research group with other supersaturated Si samples [28]. Hall mobility values are in the order of 0.01–2 cm^{2}V^{-1}s^{-1}, also in agreement with our previous measurements. The sheet concentration values are obtained from the sheet conductance and the Hall mobility, as explained in Sect. 2.4.4, and are represented in the next figure:

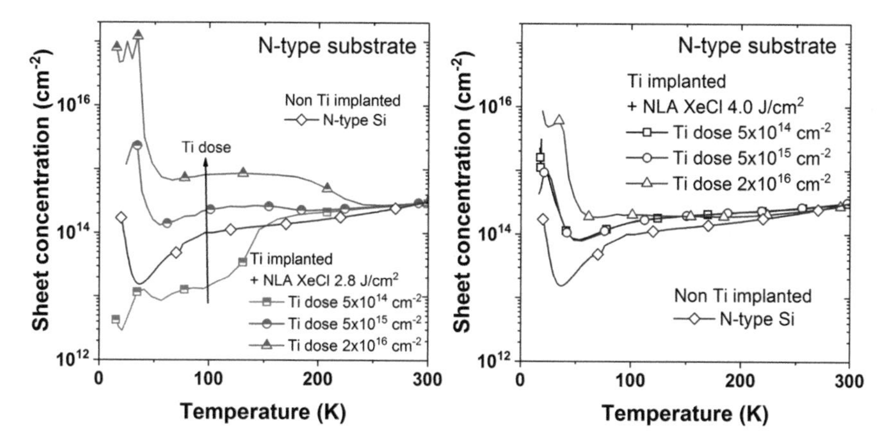

Fig. 4.20 Sheet concentration as a function of the temperature for Ti implanted samples at three different doses and a non-Ti-implanted reference. Ti implanted samples shown in the left figure were laser annealed using 2.8 J/cm^2, while the ones represented on the right were laser annealed using 4.0 J/cm^2

A similar behaviour is observed in Fig. 4.20 left: there are two transitions in the already mentioned range of temperatures. Above the high temperature transition the sheet concentration of Ti implanted samples is similar to the reference. For temperatures lower than this transition, the sheet concentration exhibits different values, increasing along with the Ti dose. For temperatures lower than the low temperature transition, the sheet concentration increases (except in the lower Ti dose) to levels close to the Ti implanted dose.

The sheet concentration dependence on the temperature for the Ti implanted samples, laser annealed with 4.0 J/cm^2 of laser fluence is similar to the reference except at low temperatures, below around 100 K. All the Ti implanted samples in Fig. 4.20 right show higher sheet concentrations than the reference below 100 K, with values very close to the Ti implanted dose at very low temperatures (around 20 K).

If we compare samples with the same Ti dose but annealed with different laser fluences, it seems that higher laser fluences lead to lower sheet concentrations at low temperatures. These results are in line to what was observed by SIMS: higher laser fluences tend to push out more Ti atoms from the Si lattice. The case of the lower dose shall not be considered as the sample laser annealed at 2.8 J/cm^2 showed high noise levels below 140 K.

4.3.4.2 Beyond the Bilayer Model

Examining as a whole the sheet conductance, Hall mobility and sheet concentration curves of the n-type Ti supersaturated Si samples we come the next observations:

- Sheet conductance values of any Ti implanted sample that are lower than the reference point out to an electrical decoupling process of the substrate from the TIL, similar to what was described for Ti implanted samples annealed with the KrF laser [33].
- Some samples exhibited a change in the sign of the Hall voltage, going from a conduction dominated by electrons at high temperatures to a conduction dominated by holes at very low temperatures. This observation is consistent to what was described by Olea et al. [33] with Ti supersaturated Si and Garcia-Hemme et al. [34] with V supersaturated Si samples. This further verifies that the samples fabricated with the XeCl laser behave similarly to what has been fabricated before in our research group with the KrF laser.
- Sheet concentration values at very low temperatures (i.e. lower than 30 K) are in relative accordance to the implanted Ti doses, as previously measured in our previous works.
- There are two transition regimes, as opposed to the single transition found on samples previously analysed in our research group.
- The first transition regime seems to happen at the same temperature for all the samples, around 60 K. This first transition regime is better represented in the Hall mobility (Fig. 4.18) or sheet concentration curves (Fig. 4.20), rather than in the sheet conductance figure.
- The second transition regime is found at relatively high temperatures (above 100 K), and its position depends on both the Ti dose and the laser fluence. An increase in the Ti dose would displace the second transition regime to higher temperatures (observation in accordance to what Olea et al. [33] measured with Ti supersaturated Si with the KrF laser). Furthermore, increasing the laser fluence displaces the second transition regime to lower temperatures, as evidenced in Figs. 4.17, 4.18 and 4.19. This observation is in accordance to what was observed by Garcia-Hemme et al. [34] with V supersaturated Si.

In a first approach, we applied the bilayer model [33] in the samples fabricated in this thesis, with limited success. The model accurately described the very low temperature regime, the second transition and the high temperature regime, but failed to reproduce the low temperature regime, along with the first transition. After further examination of the available experimental data, we developed a new model that would contemplate two different decoupling functions, labelled as the "trilayer model", in order to describe the two transition regimes, found in the experimental data. This model would use two different F function with different temperature behaviours. In the new model proposed in this thesis, the extra layer participates in the conduction mechanisms, introducing also an extra interface, where another current blocking mechanism could take place. Therefore, the TIL would be divided in two layers. Each of the layers would have a different majority carrier type, as suggested from the Hall voltage sign change in the experimental data. Hence, we labelled the layers as "TIL, N" for electrons and "TIL, P" when holes are the majority carriers. The nature of this extra layer will be discussed later. The new model could be graphically represented similarly to Fig. D.1 with an extra layer (Fig. 4.21).

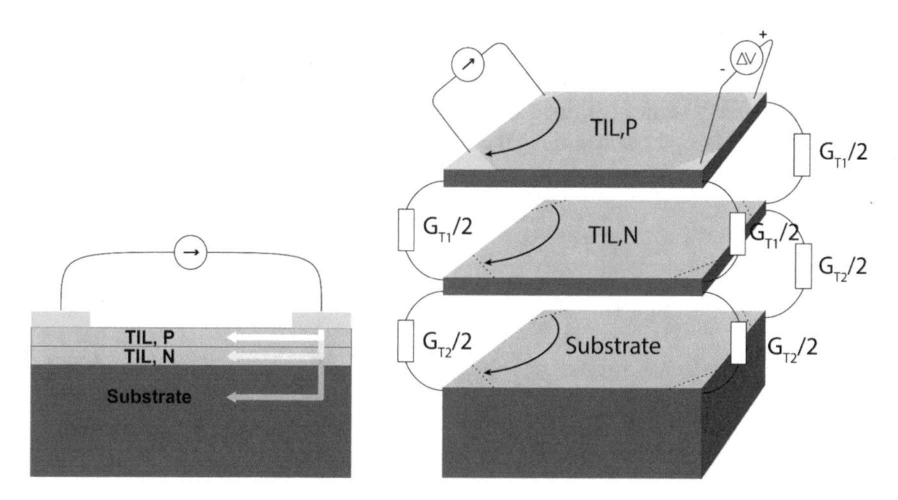

Fig. 4.21 Schematic of the trilayer model, in cross-section (left) and in exploded view (right)

In order to mathematically implement the extra layer and the extra interface into the model, we performed a nested approach: first, we fit the first transition regime of the experimental data with the decoupling function F_1 between the upper layer (p-type, denoted TIL, P) and the lower layer (n-type, denoted TIL, N) of the TIL. According to the bilayer model, we obtain an effective value of sheet conductance G_{SH1}, Hall mobility μ_{EFF1} and sheet concentration n_{EFF1} that contemplates the decoupling mechanism via a transversal conductance G_{T1} and accounts for the conduction in both layers.

$$G_{SH1} = \frac{\left(G_{TIL,P}+G_{TIL,N}F_1\right)^2}{G_{TIL,P}+G_{TIL,N}F_1^2} \tag{4.4}$$

$$F_1 = \frac{G_{T1}}{G_{T1}+\frac{G_{TIL,N}}{\alpha}} \tag{4.5}$$

$$G_{T1}(T) = G_{01}e^{-\frac{\Delta E_1}{kT}} \tag{4.6}$$

$$G_{01} = G_{001}e^{-\frac{\Delta E_1}{kT_{MN1}}} \tag{4.7}$$

$$\mu_{EFF1} = \frac{\mu_{TIL,P}G_{TIL,P}+\mu_{TIL,N}G_{TIL,N}F_1^2}{G_{TIL,P}+G_{TIL,N}F_1^2} \tag{4.8}$$

$$n_{EFF1} = \frac{G_{SH1}}{q\mu_{\mathrm{EFF1}}} = \frac{\left(G_{TIL,P}+G_{TIL,N}F_1\right)^2}{q\left(\mu_{TIL,P}G_{TIL,P}+\mu_{TIL,N}G_{TIL,N}F_1^2\right)} \tag{4.9}$$

After applying the bilayer model to the two components of the TIL, we apply a second iteration, where we apply the same equations of the bilayer model, but considering a different decoupling function, F_2, with a different transversal conductance G_{T2}, between the effective values of sheet conductance, sheet concentration and Hall mobility values obtained from the two components of the TIL and those of the substrate. Here, the substrate is the layer to be isolated from the TIL:

$$G_{SH2} = \frac{(G_{SH1}+G_{SUB}F_2)^2}{G_{SH1}+G_{SUB}F_2^2} \tag{4.10}$$

$$F_2 = \frac{G_{T2}}{G_{T2}+\frac{G_{SUB}}{\alpha}} \tag{4.11}$$

$$G_{T2}(T) = G_{02}e^{-\frac{\Delta E_2}{kT}} \tag{4.12}$$

$$G_{02} = G_{002}e^{-\frac{\Delta E_2}{kT_{MN2}}} \tag{4.13}$$

$$\mu_{EFF2} = \frac{\mu_{EFF1}G_{SH1}+\mu_{SUB}G_{SUB}F_2^2}{G_{SH1}+G_{SUB}F_2^2} \tag{4.14}$$

$$n_{EFF2} = \frac{G_{SH2}}{q\mu_{\mathrm{EFF2}}} = \frac{(G_{SH1}+G_{SUB}F_2)^2}{q\left(\mu_{EFF1}G_{SH1}+\mu_{SUB}G_{SUB}F_2^2\right)} \tag{4.15}$$

Therefore, the function F_1 relates to the interface between the p-type and the n-type layer of the TIL, while the function F_2 relates to the interface between the n-type part of the TIL and the substrate. Under these assumptions, the conduction through the trilayer could be described as follows:

- Very low temperature regime. Both decoupling functions are close to zero. The electrical current only flows through the upper layer, hence giving the p-type nature of the Hall voltage.
- First transition regime: the value of F_1 increases with temperature. $F_2 = 0$. The n-type part of the TIL starts to couple with the p-type part of the TIL. The substrate is decoupled.
- Low temperature regime: both upper layers are connected in parallel, and the measured parameters are the average of both. The substrate is still electrically isolated from the two upper layers. $F_1 \approx 1$, $F_2 = 0$.
- Second transition regime: the value of F_2 increases with temperature. The substrate starts to couple to the two upper layers.
- High temperature regime: the three layers are connected in parallel. F_1 and F_2 are equal to the unity.

We will use the curve of the sample implanted with a Ti dose of 5×10^{15} cm^{-2}, laser annealed with 2.8 J/cm^2 as a representative curve to evaluate the application of the bilayer and trilayer models. Two main assumptions have been done before fitting the model to the experimental data: Hall mobility and sheet concentration of each TIL has been modelled as constant with temperature, to account for a metallic-like behaviour [28]. These assumptions have been done in both n- and p-type parts of the TIL in the case of the trilayer model, and in the only component of the TIL used for the

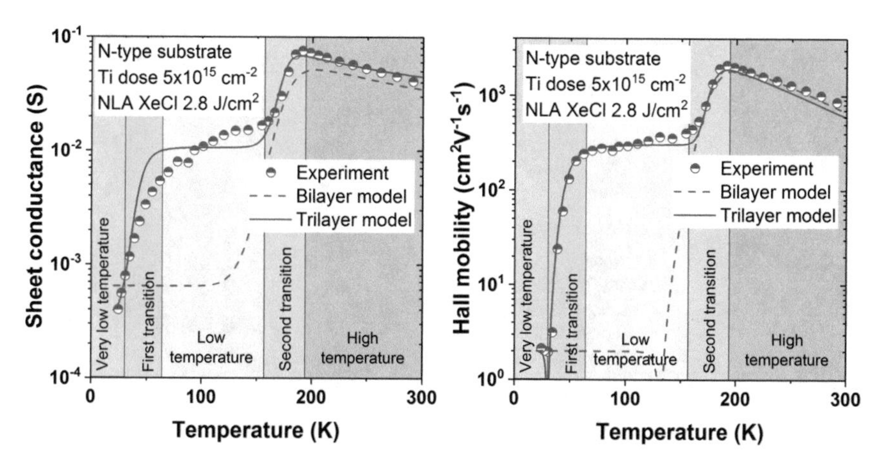

Fig. 4.22 Sheet conductance (left) and Hall mobility (right) as a function of the temperature for a sample Ti implanted with a total dose of 5×10^{15} cm^{-2}, subsequently laser annealed with a laser fluence of 2.8 J/cm^2. The points indicate the experimental data. The dashed line represents the bilayer model (as formulated by Olea et al.), while the solid line represents the trilayer model (the new model developed in this thesis). The five different regimes are indicated in both graphs as shaded areas along with their respective names

bilayer model (as it was previously assumed by Olea et al. and Garcia-Hemme et al.). The sign of each carrier has been taken into account in the formulation (Fig. 4.22).

The quality of the fitting is considerably better on the trilayer model, as the bilayer model has been found insufficient to fit the low temperature region. Also, Hall mobility curves was better fitted by the model than the sheet conductance. In the latter, the first transition seems less abrupt, which was not properly fitted in the trilayer model. After evaluating the differences between the two models, we show the fitting of trilayer model for all the samples, in the sheet conductance and Hall mobility curves (Fig. 4.23).

The trilayer model accurately reproduced the experimental curve of the six samples analysed, taking into account that the sample with a Ti dose of 5×10^{14} cm^{-2}, laser annealed at 2.8 J/cm^2 is noisy below 140 K. As described before, two out of six samples showed a change in the sign of the Hall voltage. These three samples (the ones shown in Fig. 4.19) showed the previously described two transitions. The other three samples only showed one transition. In the trilayer model, we assumed for the latter samples that the p-type and the n-type part of the TIL were in parallel, or with activation energies (see Eqs. 4.8 and 4.14) close to zero.

The bilayer model uses four parameters: the transversal conductance at 0 K G_0, the activation energy ΔE and the Hall mobility and sheet concentration of the TIL, μ_{TIL} and n_{TIL} respectively. Since the trilayer model uses an extra layer, it doubles the number of needed parameters, up to eight. As in the bilayer model, the number of parameters can be reduced if we consider that the Meyer-Neldel rule applies to these samples, where the transversal conductance at 0 K and the activation energy are linked by an Arrhenius law (Eqs. 4.9 and 4.15). This would reduce the parameters to three

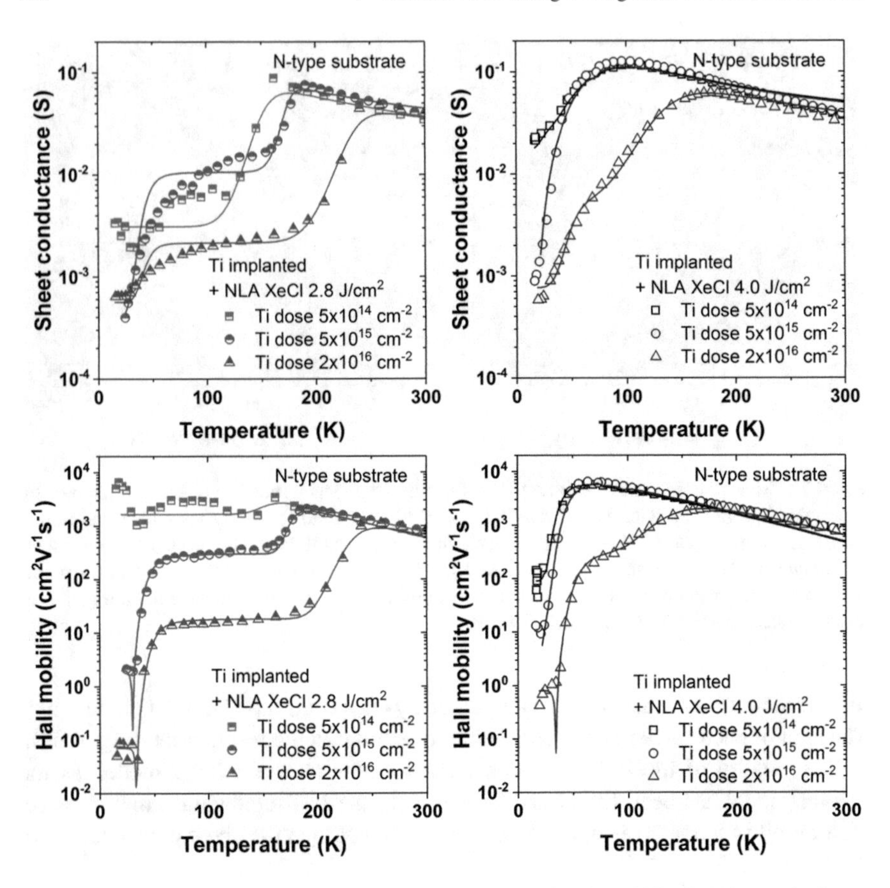

Fig. 4.23 Sheet conductance in Y-log scale (upper row) and Hall mobility in Y-log scale (lower row) as a function of the temperature for Ti implanted samples based on an n-type substrate. Left column shows a set of three samples laser annealed at the same laser fluence of 2.8 J/cm^2, while the right column shows three samples laser annealed at a laser fluence of 4.0 J/cm^2. In all graphs, symbols represent the experimental data and the solid lines the fitting to the trilayer model

instead of four. This law was proven to be applicable on V supersaturated Si samples by Garcia-Hemme et al. [34]. However, we did not consider the Meyer-Neldel rule in the first approach, leaving eight parameters in the trilayer model. We will evaluate the suitability of the Meyer-Neldel rule to our experimental fitting parameters later. Table 4.2 shows the fitting parameters of the six samples.

There are two trends in the first (low temperature) and second (i.e. high temperature) transitions: both the transversal conductance and the activation energy decrease when increasing the laser fluence, as it was observed by Garcia-Hemme et al. [34], although they considered only one transversal conductance. This can be better observed by representing the transversal conductance in log Y-scale as a function of the activation energy (Fig. 4.24).

Table 4.2 Transversal conductance at 0 K and activation energy parameters for the low temperature transition (sub-index 1) and the high temperature transition (sub-index 2) of Ti implanted samples, laser annealed with the XeCl laser using an n-type Si substrate

Ti dose (cm^{-2})	Laser fluence (J/cm^2)	G_{01} (S)	ΔE_1 (eV)	G_{02} (S)	ΔE_2 (eV)
5×10^{14}	2.8	0.1	0	100	0.123
	4.0	0.1	0	0.25	0.010
5×10^{15}	2.8	0.1	0.180	10^8	0.335
	4.0	5×10^{-3}	0.007	1.0	0.018
2×10^{16}	2.8	0.2	0.250	1500	0.251
	4.0	0.1	0.220	0.5	0.048

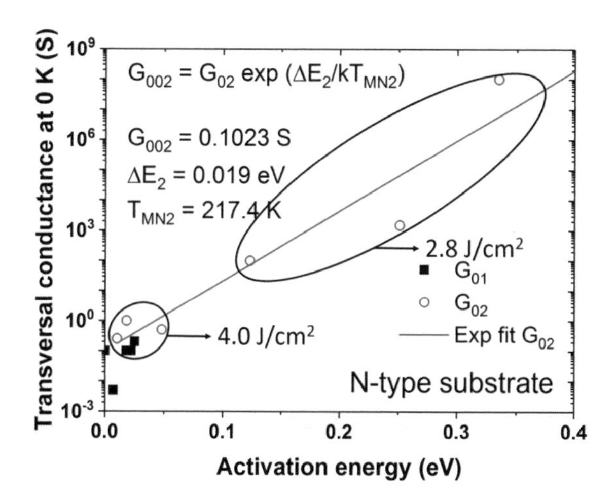

Fig. 4.24 Experimental fitting to the Meyer-Neldel rule for Ti implanted samples based on n-type Si substrates

The activation energy obtained from the Meyer-Neldel rule, 19 meV, is very close to the value previously published in our research group by Garcia-Hemme et al. [34], where they found an activation energy of 22 meV for V supersaturated Si and 15 meV for Ti supersaturated Si. However, the transversal conductance at 0 K was different in our case, five times higher than the value reported in the cited reference for V atoms in Si.

Finally, we show the values of Hall mobility and sheet concentration used to fit the curves of Ti implanted samples using an n-type Si substrate (Table 4.3).

From the available data, we observe how the sheet concentration of the p-type part of the TIL increases with the Ti dose, as expected, while its concentration decreases with increasing laser fluence, which is in accordance with what was observed with SIMS measurements, due to the snow-plow effect [35]. The Hall mobility of both layers in samples annealed at 4.0 J/cm^2 increases along with the Ti dose, while the opposite trend is found on the samples annealed at 2.8 J/cm^2. The rest of parameters do not show any general trend, which may require further measurements of more samples to obtain better statistics.

Table 4.3 Hall mobility and sheet concentration fitting parameters for the p-type (surficial) layer and the n-type (intermediate) layer of Ti implanted samples, laser annealed with the XeCl laser using an n-type Si substrate

Ti dose (cm^{-2})	Laser fluence (J/cm^2)	$\mu_{TIL,P}$ ($cm^2V^{-1}s^{-1}$)	p_{TIL} (cm^{-2})	$\mu_{TIL,n}$ ($cm^2V^{-1}s^{-1}$)	n_{TIL} (cm^{-2})
5×10^{14}	2.8	1.0	1.5×10^{15}	1300	1.8×10^{13}
	4.0	0.07	5.0×10^{14}	90	1.2×10^{15}
5×10^{15}	2.8	2.0	1.2×10^{15}	320	2.0×10^{14}
	4.0	0.3	10^{15}	100	10^{15}
2×10^{16}	2.8	0.07	5.0×10^{16}	25	4.0×10^{14}
	4.0	0.8	6.0×10^{15}	400	8.0×10^{13}

Discussion on the trilayer model

The four-probe measurements suggested that two decoupling mechanisms could be happening at different temperatures on Ti implanted samples based on n-type substrates, annealed using a XeCl laser with a pulse duration of around 150 ns. We developed a model, based on the already described bilayer model, which we labelled as the trilayer model. The trilayer model fitted better the experimental results, being able to predict the shape of the sheet conductance and Hall mobility experimental curves.

The experimental curves of sheet conductance, Hall mobility and sheet concentration showed two differenced behaviours: some samples exhibited one transition regime while others showed two different transitions at different temperatures. If we examine again the compositional and structural measurements (SIMS, HRTEM, Raman spectroscopy, see Sect. 4.3.1), we may find a possible explanation for this dual behaviour.

From HRTEM and SIMS measurements mostly, we concluded that the laser process did not anneal deep enough to melt past the deeper Ti atoms in some samples. In particular, all the samples laser annealed at 2.8 J/cm^2 and the sample with a Ti dose of 2×10^{16} $cm^{-2}$and laser annealed at 4.0 J/cm^2. If we examine again the sheet conductance, Hall mobility and sheet concentration measurements of these samples, we observe that they are precisely the ones that show the double coupling mechanism, associated to the trilayer structure. Inside the TIL there seems to be two different sub-layers: one laser annealed, which possibly melted and solidified after the NLA process, and a second one that was not melted, but heated up instead, due to the proximity to the melted region. These two layers may exhibit different conduction mechanisms, as the Ti concentration and the density of defects may be different.

The trilayer model, proposed in this thesis, is a first step towards the understanding of the conduction mechanisms on samples Ti implanted with high energies, which result in thick Ti implanted layers, laser annealed with longer pulses. The model accurately fitted the experimental data for all the samples and the first results (as sheet concentration of the layers or the Hall mobility values) are consistent with the fabrication parameters. The extracted parameters are also in agreement with those previously obtained in previous studies in our research group. We expect to measure the remaining samples aiming to further refine and test the model.

Finally, the observations done in this sub-section, related to the nature of the blocking mechanism, agree with the results of Garcia-Hemme et al. [34]. The activation energy observed in the four-probe measurements of what we have called the second transition could be related to the height of the barrier caused by the implantation tails. Increasing the laser fluence increases the melted depth, which decreases the amount of Ti atoms left non-melted after the NLA. Thinner non-melted layers may lead to lower values of the electrical barrier, consistent with the extracted values (see Table 4.2).

4.3.4.3 P-Type Si Substrates

The p-type Si substrate used in this chapter has the code P2, according to Table 2.1. We measured five samples based on them: three at a fixed laser fluence of 4.0 J/cm^2, sweeping the Ti dose, and a set of three samples where we fixed the Ti dose at 5×10^{15} cm^{-2} and swept the laser fluence. We also measured a p-type reference substrate with no Ti implantation, but back implanted with B to produce a BSF.

In Fig. 4.25 we have also represented the fitting to the trilayer model, already explained in previous sections. These samples, based on p-type substrates, show similar behaviour to their n-type counterparts: some of them exhibit one transition, while others present two transitions. The first transition seems to be independent on the Ti dose, as we observed with n-type substrates, in the order of 70 K. The

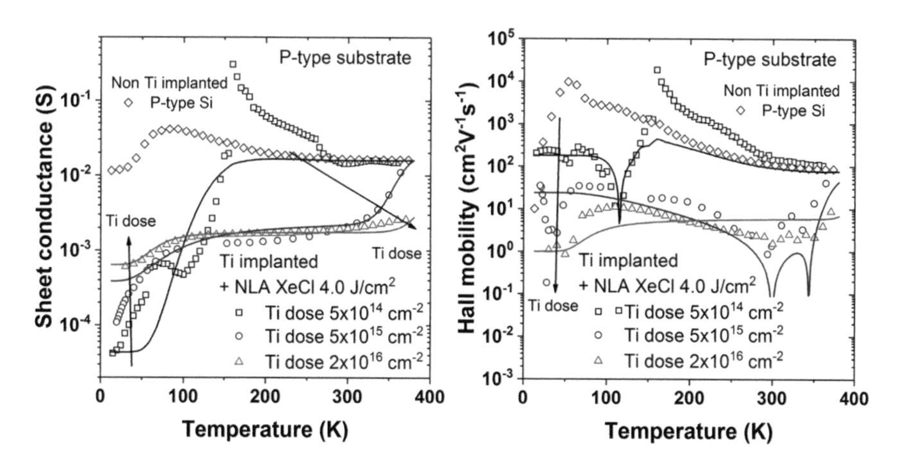

Fig. 4.25 Sheet conductance (left) and Hall mobility (right) of implanted samples at different Ti doses, at a fixed laser fluence of 4.0 J/cm^2. The substrate is p-type, with a boron implantation on the backside to produce a BSF. Scatter points represent the experimental data, while solid lines are the trilayer model fitting to the experimental data

second transition happens at considerably higher temperatures than in the case of n-type substrates, and it follows the same behaviour: increasing the Ti dose moves the second transition to higher temperatures. Here, the lower dose exhibits a sharp decay at around 160 K, the middle dose at 360 K and for the higher dose we did not observe it in the examined range of temperatures, suggesting it could be happening at temperatures higher than 380 K. In the sheet conductance curves, we observe that in the range of very low temperatures the sheet conductance increases along with the Ti dose. In the case of Hall mobility, the mobility decreases with increasing Ti dose, which would lead to higher sheet concentrations when increasing the Ti dose, as expected.

The measurement of the Hall effect was not straight-forward on most p-type samples, mainly due to the low signal-to-noise ratios, at temperatures lower than the second transition. The noise in this region made it difficult to interpret the results, with several changes in the sign of the Hall voltage.

When applying the trilayer model, we assumed the same model used on Ti supersaturated layers on n-type substrates: the p-type part of the TIL is in direct contact with the electrodes, while the n-type part of the TIL is in contact with the substrate. Nor the bilayer not the trilayer model adequately fitted the experimental data on p-type substrates in the whole temperature range, or at least not as accurate as we saw in Ti supersaturated n-type Si samples.

After discussing the effect of the Ti dose on the conduction properties, we switch to the effect of the laser fluence at a fixed Ti dose of 5×10^{15} cm^{-2}, showing also the fitting to the trilayer model. The sheet conductance curves (Fig. 4.26 left) show the strong influence of the laser process, in the whole range. In general, increasing the laser fluence increases the conductivity as well, which is consistent with a better

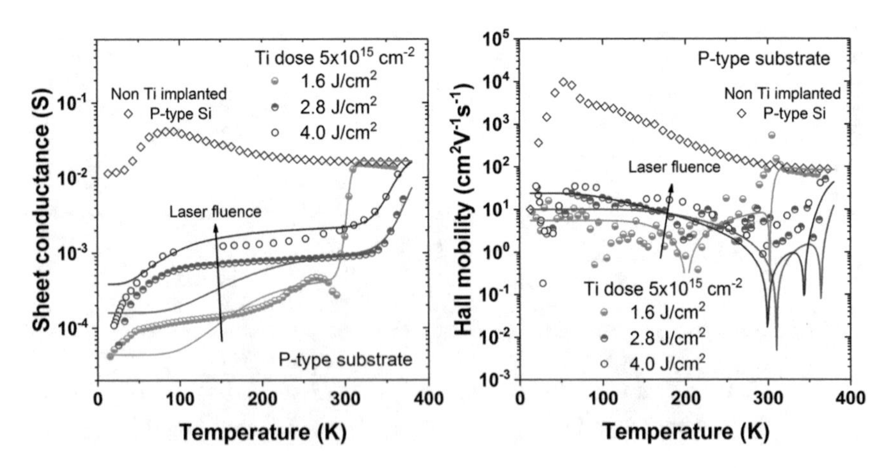

Fig. 4.26 Sheet conductance (left) and Hall mobility (right) of implanted samples at a fixed Ti dose of 5×10^{15} cm^{-2}, annealed at different laser fluences. The substrate is p-type, with a boron implantation on the backside to produce a BSF. Scatter points represent the experimental data, while solid lines are the trilayer model fitting to the experimental data

crystal quality in the implanted layer. As it happened in the previous set of samples, the Hall effect measurements provided results with poor signal-to-noise ratios. The Hall mobility curves show high dispersion for temperatures lower than the second transition. It is worth mentioning that the three samples shown in Fig. 4.26 show conduction dominated by electrons at room temperature.

We fitted the experimental data to the trilayer model (shown as solid lines in Fig. 4.26). We chose fitting parameters that could minimise the error in both the sheet conductance and the Hall mobility curves. The fitting parameters are shown below, in a similar way as we did with the n-type Si substrates (Table 4.4).

Later, we applied the Meyer-Neldel rule on those samples (Fig. 4.27).

The activation energy obtained from the Meyer-Neldel rule, 26 meV, is still close to the value previously published in our research group by Garcia-Hemme et al. [34], where they found an activation energy of 22 meV for V supersaturated Si. The transversal conductance at 0 K is considerably lower (two orders of magnitude) than the value reported by the same authors. The values are different from what we found in n-type substrates: the activation energy is 7 meV higher, and the transversal

Table 4.4 Transversal conductance at 0 K and activation energy parameters for the low temperature transition (sub-index 1) and the high temperature transition (sub-index 2) of Ti implanted samples, laser annealed with the XeCl laser using a p-type Si substrate

Ti dose (cm^{-2})	Laser fluence (J/cm^2)	G_{01} (S)	ΔE_1 (eV)	G_{02} (S)	ΔE_2 (eV)
5×10^{14}	4.0	0.01	0.002	0.01	0.039
5×10^{15}	1.6	0.001	0.052	10^{25}	1.65
	2.8	0.001	0.020	10^{9}	0.87
	4.0	0.001	0.015	10^{8}	0.75
2×10^{16}	4.0	0.005	0.020	100	0.42

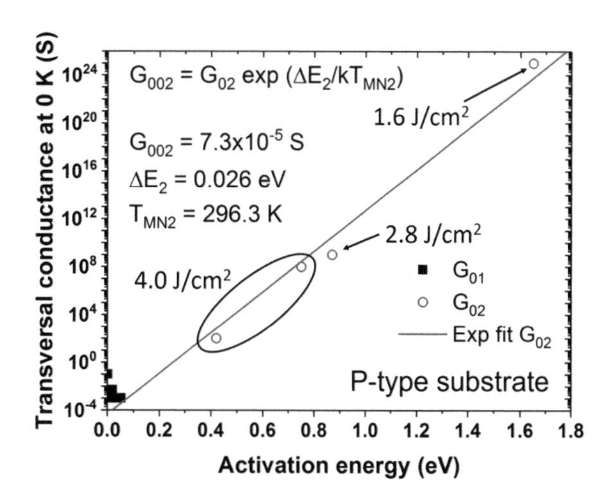

Fig. 4.27 Experimental fitting to the Meyer-Neldel rule for Ti implanted samples based on p-type Si substrates

conductance at 0 K is four orders of magnitude lower. As we measured with n-type substrates, lower laser fluences are showing the higher activation energies.

The study of the conduction mechanisms on Ti implanted layers over p-type Si substrates is a novelty of this thesis in our research group. The results displayed in this sub-section show a first approximation to the modelling of the conduction processes of TIL atop p-type Si substrates, where some similarities and some differences were found when compared to n-type Si substrates. The nature of the second transition regime, associated to the decoupling function F_2, seems to follow the Meyer-Neldel rule, as in TIL over n-type substrates, and in line with previous works in our research group [34]. However, the measurement of the Hall effect has been found to be challenging, obtaining results close to the noise level of the technique, which halted further analysis of the samples. We expect to extend our research of TIL over p-type substrates to better understand the nature of the junction between them, by measuring more samples and using AC Hall effect measurements [36].

4.3.4.4 Conclusions on Four-Probe Measurements

There are some similarities and differences between Ti implanted layers over n-type or p-type silicon substrates. The analysis of the n-type substrates allowed us to check that similar conduction mechanisms were present in the samples annealed with the KrF and XeCl lasers. However, with the XeCl laser we found two different decoupling mechanisms, which led to the development of the trilayer model to better describe the experimental measurements, although it still needs further improvement. After, we measured and analysed the electrical properties of TIL over p-type Si substrates, which had been barely studied before in our research group. The sheet conductance values led to very low conductivities at room temperature, which was associated to the decoupling of the substrate from the TIL. Garcia-Hemme linked the decoupling mechanisms on n-type substrates, supersaturated with transition metals, to the sub-bandgap photoresponse [27]. In particular, it was found that the photoresponse increased when the temperature was close to the transition region; the light could be lowering the barrier between the TIL and the substrate, effectively changing the sheet conductance. Following this idea, p-type substrates could exhibit better photoresponse at room temperature than n-type substrates, as most of them are decoupled from the substrate at room temperature. The trilayer model allowed us to check that the Meyer-Neldel rule applied to both n-type and p-type substrates, suggesting that the origin of the electrical barrier would be related to the implantation tails between the TIL and the substrate, as pointed out by Garcia-Hemme et al. [34].

Therefore, the analysis performed in this sub-section is useful to develop the design rules for future sample fabrication: increasing the Ti dose moves the second transition regime to higher temperatures, as well as lowering the laser fluence produces the same effect. By appropriately choosing both parameters, it would be possible to design a material with the second transition regime close to room temperature, with good monocrystal quality (by looking at the average value of mobility in the low temperature region), which would result in photodetectors in the SWIR with improved photoresponse, according to the findings of Garcia-Hemme et al. [27].

4.4 Planar Photodiode Fabrication

The main goal of the thesis is to develop a material, based on Si substrates, that could extend the responsivity of Si at photon energies lower than the bandgap at sufficient low cost, in order to transfer the results to the industry. The devices fabricated in this section are a first prototype used to check if the Ti supersaturated material obtained using the XeCl laser is suitable for that matter. The prototypes are based on planar transversal photodiodes, done by contacting the TIL and the substrate on the frontside and the back surface of the wafer, respectively. To speed up the fabrication process, we fabricated the simplest photodiodes: square-shaped samples of $9 \times 9\ \text{mm}^2$ without any passivation layer, just the deposition of the front and back electrodes by means of e-beam evaporation through a hard mask. We used a metallic stack 50 nm + 100 nm thick composed of Ti (in contact with Si) + Al (in contact with air), respectively. Silver paint was used to connect the samples to chip carriers. The chip carriers used for the photodiode characterisation are not compatible with the cryostat. Therefore, all measurements were performed at room temperature (Fig. 4.28).

Only p-type substrates were used in this section for photodiode fabrication. The material characterisation in the previous sub-section pointed to p-type substrates as best candidates to increase the photoresponse at room temperature due to the position of the second transition regimes. Besides, STMicroelectronics works mainly on p-type substrates for CMOS Imaging Sensors; using n-type substrates was not a possibility for further integration, not at least for the first prototypes (see Chapter 5).

Along with the Ti implanted samples, we fabricated a reference N^+P diode following the P and B implantation recipes (for frontside and backside implantation respectively) described in Sect. 3.8.1.

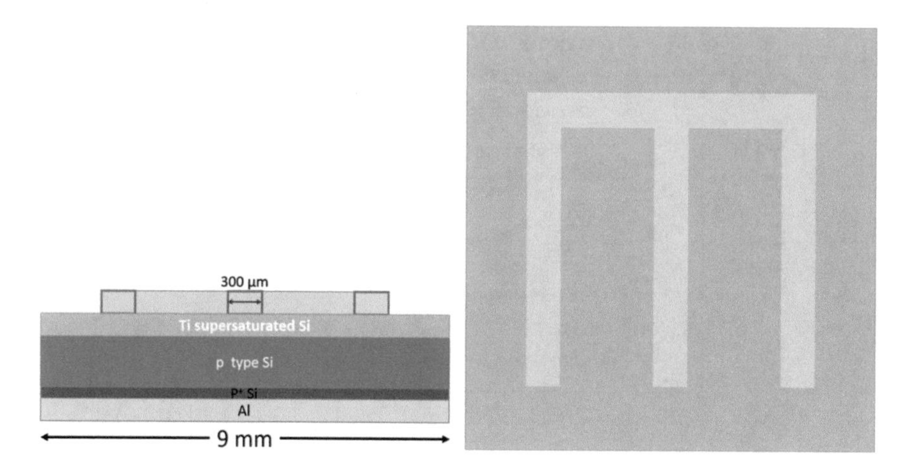

Fig. 4.28 Schematic of the photodiode structure. Left: cross-section. Right: top-view

4.5 Planar Photodiode Characterisation

We characterised the photodiodes using two techniques: current-voltage measurements and photoresponse measurements for electric and optoelectronic characterisation.

4.5.1 Electrical Characterisation

The electrical characterisation, by means of current-voltage characteristics, was performed using the instruments described in Sect. 2.4.3. The photodiodes were polarised by sweeping the bias on the back electrode (the p-type substrate), while keeping the top contact (the TIL or the P implanted layer in the case of the N^+P reference diode) grounded. The curves are represented in such a way that direct current is on the right.

In Fig. 4.29, we represent the IV curves for Ti implanted samples with different Ti doses, along with the N^+P reference diode. Scatter points show the experimental data, while solid lines represent the fitting to a real diode model. The samples show rectification between three to five orders of magnitude. Direct current corresponds to a positive bias applied to the substrate, which would imply that the TIL would behave as an n-type layer. The van der Pauw measurements shown in the previous

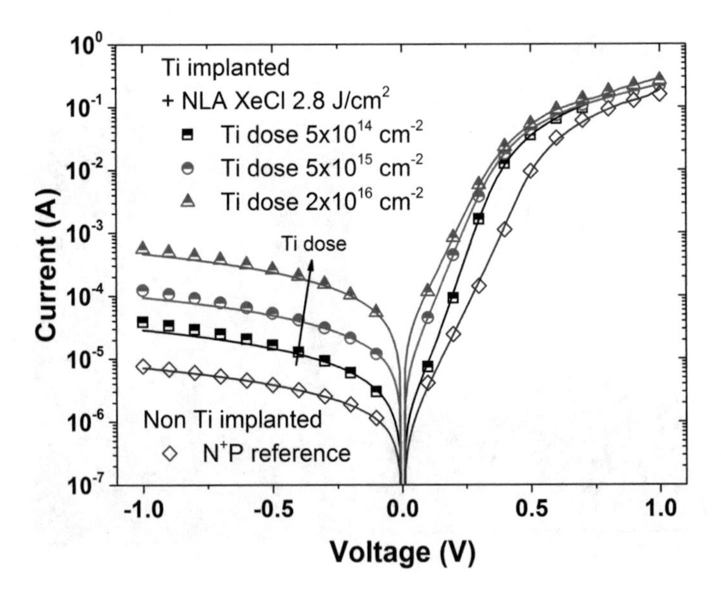

Fig. 4.29 Current-voltage curves of Ti implanted Si samples at different Ti doses, annealed at a fixed laser fluence of 2.8 J/cm^2 using the XeCl laser. The absolute value of the current is represented. Lines indicate the fitting to a real diode model

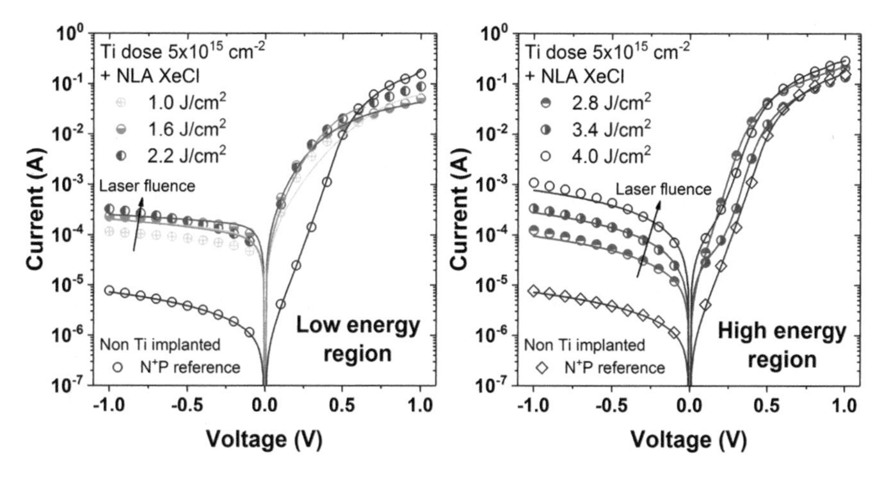

Fig. 4.30 Current-voltage curves of Ti implanted Si samples at a fixed dose of 5×10^{15} cm^{-2}, annealed at different laser fluences using the XeCl laser. Left figure shows three samples with fluences lower or equal than 2.2 J/cm^2 (the low energy region), and right figure shows three samples with fluences equal or higher than 2.8 J/cm^2. The absolute value of the current is represented. Lines indicate the fitting to a real diode model

subsection also show conduction dominated by electrons at room temperature in the same samples. Both direct and reverse current increase along with the Ti dose in the whole bias range studied, showing higher currents than the reference diode as well. The reverse current of the three Ti implanted samples go almost parallel, but it is not the case of the direct part of the IV, where we observe differences in the slope.

Next, we analyse the effect of the NLA process on the IV curves, at a fixed Ti dose of 5×10^{15} cm^{-2}.

We found two regimes when analysing the IV curves, as a function of the laser fluence, already described in previous sections of this chapter: low energy regime for fluences lower or equal than 2.2 J/cm^2 (represented in Fig. 4.30 left) and the high energy regime for fluences equal or higher than 2.8 J/cm^2 (Fig. 4.30 right). In the low energy regime, we observe IV curves which remind of a resistive behaviour in direct current. The slope of the curve in voltages close to +0.4 V is low, indicating high diode ideality factors, and the reverse current show little dependence on the reverse voltage. In reverse, we observe that increasing the laser fluence increases the reverse current at −1 V. In the high energy regime, the reverse current exhibits a resistive behaviour, similar to shunting effects. In this region, we observe a clear influence of the laser fluence in the whole IV curve: higher fluences lead to higher currents. The direct current part of the IV curves shows an exponential behaviour up to 0.5 V, where we observe the effect of a series resistance. The sample annealed at 2.8 J/cm^2 of laser fluence show the highest slope, while the sample annealed at 4.0 J/cm^2 show the highest current values at 1 V.

Given the rectifying behaviour observed on the fabricated photodiodes, we applied a real diode model to the experimental data, in order to obtain more information about the conduction processes (already shown in Figs. 4.29 and 4.30 as solid lines). The quality of the fitting to a real diode model is good for all the samples, even in reverse current. The extracted parameters for all the samples are represented in the next graph as a function of the laser fluence, for the three Ti doses analysed in this chapter.

In Fig. 4.31 we represented the fitting parameters of the reference N^+P diode to a double diode model, as the model with only one diode proved to be insufficient. Therefore, we show two diode ideality factors and two dark saturation currents. The values are: $n_1 = 1$, $n_2 = 2$; $I_{01} = 5 \times 10^{-7}$ A and $I_{02} = 5.5 \times 10^{-11}$ A. The last parameter is out of the displayed vertical scale in the upper right graph of Fig. 4.31, so it is not shown.

The diode ideality factor n is strongly related to the nature of the conduction mechanisms in the sample [37]. It shows two differentiated regimes at the three Ti doses analysed: diode ideality factors higher than 2 in the low energy region (fluence lower or equal than 2.2 J/cm^2) and diode ideality factors lower than 2 in the high energy region (fluence equal or higher than 2.8 J/cm^2). The three curves with different Ti doses show that increasing the laser fluence decreases the diode ideality factor. In the high energy region, the value stabilises at values close to 1.5. In this region we observe

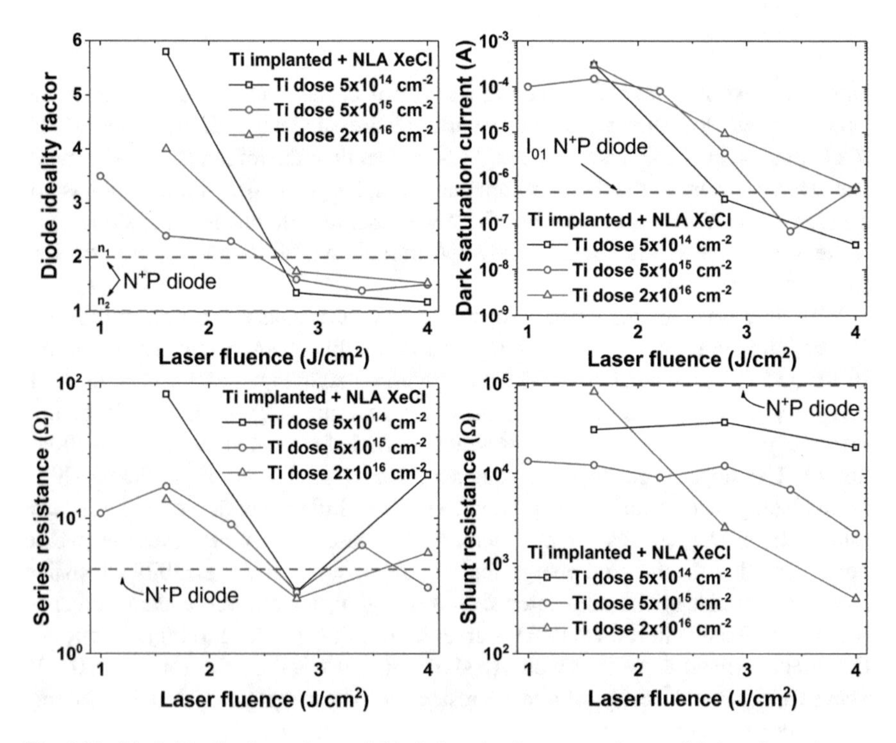

Fig. 4.31 Diode ideality factor (upper left), dark saturation current (upper right), series resistance (lower left) and shunt resistance (lower right) as a function of the laser fluence for samples implanted at three different Ti doses. Horizontal red dashed lines show the parameters of the N^+P reference. All parameters have been extracted from the fitting to a real diode model

that increasing the Ti dose increases the diode ideality factor, at a given laser fluence within this regime. Lower diode ideality factors are usually preferred, as values close to 1 relate to low recombination rates, mainly caused by minority carriers. Therefore, the high energy region is adequate to fabricate devices attending to the diode ideality factor. The experimental values for Ti implanted samples, between 1 and 2, would account for recombination of minority carriers in the neutral regions as well as in the depletion region. Finally, the high diode ideality factors found in Ti implanted samples within the low energy region could be consistent with conduction through highly defective (amorphous or nanocrystalline) layers, where tunnel conduction mechanisms could be expected.

In general, dark saturation currents as low as possible are preferred for photodetectors, as it is generally considered as a noise source, present in both dark and illumination conditions. Noise is especially affecting those scenarios with low light conditions, when the signal coming from the photodiode is low and could be totally hidden by noise. We have found experimentally that the dark saturation current is affected by both the laser fluence and the Ti implanted dose. In general, higher Ti doses give higher dark saturation current values. Besides, increasing the laser fluence decreases the dark saturation current. Both trends are consistent to the relationship between the dark saturation current and the quantity of noise coming from defects. Higher Ti dose are expected to produce more defects inside the semiconductor (more ions imply more collisions with the Si lattice), that may be more difficult to anneal at a given laser fluence. The same way, increasing the laser fluence generally decreases the amount of remaining defects after the ion implantation. Higher laser fluences also imply that the laser melts deeper into the sample, increasing the chance of annealing the end of range defects, which have been reported to be more difficult to anneal [21]. The analysis of the dark saturation current with respect to the laser fluence points out again to the regime of high energy to produce photodiodes with less noise sources and higher signal-to-noise ratios.

Next, we analyse the series resistance, where values as low as possible are desired. The series resistance values found when fitting the experimental data to a real diode model show that in general, higher laser fluences lower the series resistance value. The three Ti doses exhibit a minimum at 2.8 J/cm^2 of laser fluence, lower than the series resistance obtained in the N$^+$P diode. The results also point out to an inverse relationship between the Ti dose and the series resistance, where higher Ti doses lead to lower series resistance values, which is consistent with a lower resistance coming from the Ti implanted layer, as we measured using the van der Pauw configuration. Besides, in general, semiconductors with higher doping levels provide lower contact resistance values, switching from a rectifying (Schottky) contact to a tunnel-assisted ohmic contact [38]. Both options are non-exclusive and they could be happening at the same time.

The last parameter represented in Fig. 4.31 (right lower graph) is the shunt resistance. High resistance values are preferred; as higher resistances imply lower shunt and leakage currents. The shunt resistance is found to be dependent on both the laser fluence and the Ti dose. Increasing the Ti dose decreases substantially the shunt resistance. We suspect it could be due to the cellular breakdown structures, which are proven to be Ti-rich structures. Higher Ti concentrations are linked to higher conductivity (as seen in this chapter through four-probe measurements). The CBD structures could effectively act as preferred paths for the carriers, as they could lower the electrical barrier between the TIL and the substrate, especially considering that the CBD structures extend almost to the supersaturation limit. With respect to the influence of the laser fluence, we found that increasing the laser fluence decreased the shunt resistance. This fact points again to the CBD structures as the main cause. As we observed in the structural characterisation of the Ti supersaturated layer using cross-section HRTEM (Sect. 4.3.1) the length of the CBD structures increased by more than 20% when increasing the laser fluences from 2.8 up to 4.0 J/cm^2. This would mean that the distance between the supersaturated layer and the substrate would be lower, easing the leakage of carriers through the substrate. Another fact that further supports our hypothesis linking the CBD structures and the low shunt resistance values is the case of samples Ti implanted at 5×10^{14} cm^{-2}. These samples did not show any sign of CBD structures, and the shunt resistance of these samples is barely affected by the laser fluence. Finally, according solely to the shunt resistance parameters, fluences as lower as possible would be preferred due to their higher shunt resistance values.

Attending to the electrical parameters, laser fluences within the high energy regime seem better for the device performance. However, the pernicious effect of the shunt resistance at high laser fluences must be taken into account, as well as the slight increase in the series resistance at said values. Values between 2.8 and 3.4 J/cm^2 would lead to a compromise between the different parameters. In particular, at 2.8 J/cm^2 we obtained the lower series resistance. The shunt resistance is at acceptable levels at the three doses, the diode ideality factor is around 1.5 and the dark saturation current is sufficiently low, in the order of tens of nA. With respect to the best Ti dose, the samples Ti implanted at a dose of 5×10^{14} cm^{-2} would lead to better diodes according to the examined electrical parameters. However, this last affirmation confronts with the sub-bandgap absorption in this set of samples, where they showed lower absorption coefficients than other samples with higher Ti doses.

4.5.2 Photoresponse

Photoresponse is the key parameter of any photodiode, where the capacity of the device to detect light is measured. In this subsection, we analyse the External Quantum Efficiency (EQE) in the visible and the IR part of the spectra, using the two photoresponse clusters described in Sect. 2.4.5, at room temperature. Although our research is focused on the enhancement of the EQE in the sub-bandgap region (the NIR range), having a photodetector sensitive in a range of the spectrum as wide as possible is highly desirable as it would increase the number of possible applications.

First, we analyse the EQE in the UV, visible and NIR range. The effect of the Ti dose in the EQE, at a fixed laser fluence of 2.8 J/cm^2 is shown below. The Ti implanted samples show lower EQE values than the reference in the whole range. The shape of

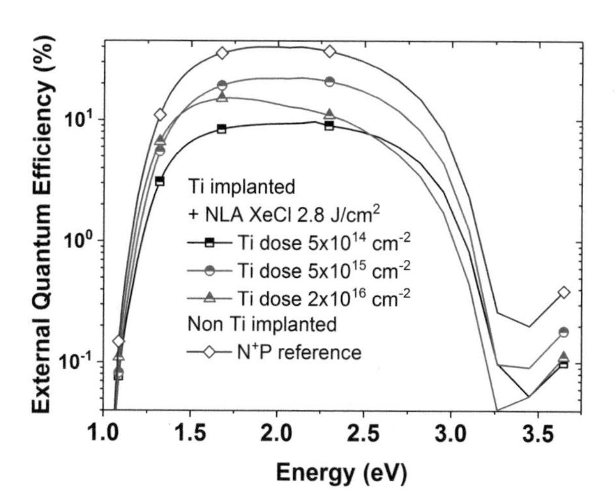

Fig. 4.32 External Quantum Efficiency as a function of the photon energy for Ti implanted samples at different doses, laser annealed at a fixed fluence of 2.8 J/cm^2. The reference N$^+$P diode is shown for reference

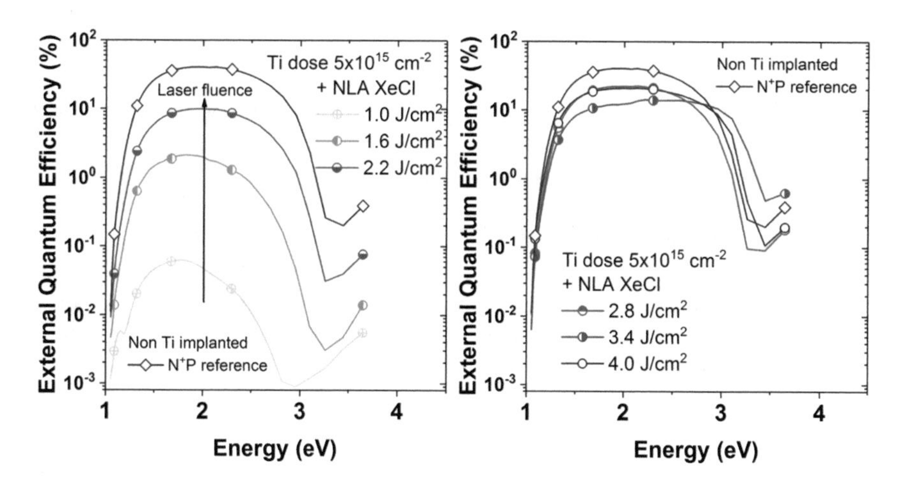

Fig. 4.33 External Quantum Efficiency as a function of the photon energy for Ti implanted at a dose of 5×10^{15} cm^{-2}, laser annealed at different laser fluences. The reference N$^+$P diode is shown for reference

the EQE curve of the lower and middle Ti dose are similar to the reference, while the sample with the highest Ti dose showed lower EQE values at high photon energies. This could indicate higher recombination rates in the latter, in regions close to the surface, where the most energetic photons are usually absorbed. The highest EQE values in most of the range are offered by the sample with the middle dose.

In the next analysis, we fix the Ti dose at 5×10^{15} cm^{-2} while we sweep the laser fluence.

Figure 4.33 left show three Ti implanted samples at increasing laser fluence in the low energy regime (fluence lower or equal than 2.2 J/cm^2). The curves show an important increase in the EQE with increasing laser fluence. This fact could be associated to a reduction of the damage coming from the ion implantation process at higher laser fluences. The Raman spectroscopy and TEM micrographs showed residual damage in these samples (TEM was only done in the sample laser annealed with 1.6 J/cm^2), which would be consistent with the behaviour observed here.

If we move to the samples annealed in the region that we have labelled as the high energy regime, the behaviour is different (see Fig. 4.33 right). The laser fluence in this region seems to affect little the EQE curve, where the slight differences could be attributable mostly to little variations between different samples. The slight influence of the laser fluence in this regime suggests that the crystal quality of the Ti implanted samples should be similar, as backed by Raman spectroscopy and cross-section HRTEM micrographs.

All the Ti implanted samples showed lower EQE values than the N$^+$P reference diode in the UV, visible and NIR part of the spectra, which is consistent with similar results reported by Olea et al. [39] on samples annealed using the KrF laser.

Although it would be desirable to obtain higher EQE values than the reference in the whole range (from UV towards the IR), the interest of supersaturated materials lies in sub-bandgap transitions, that is, for energies lower than 1.12 eV for Si samples, in the SWIR region (down to 0.4 eV).

Figure 4.34 shows the sub-bandgap EQE of Ti implanted samples at a fixed NLA process, where the same fluence of 2.8 J/cm^2 was used. All the samples, including the non Ti implanted references, show sub-bandgap photoresponse. We attribute the sub-bandgap photoresponse of the reference sample to surface defects, as we demonstrated in the previous chapter, Sect. 3.3. The samples described in this chapter were not passivated to minimise surface defects, which could explain the relatively high sub-bandgap photoresponse of the reference sample. The Ti implanted samples show higher EQE values in the whole sub-bandgap range than the reference, extending their response up to 0.45 eV (2.75 μm) in the best case at room temperature, way below the bandgap of silicon. Again, as seen in Fig. 4.32 in photon energies over the bandgap,

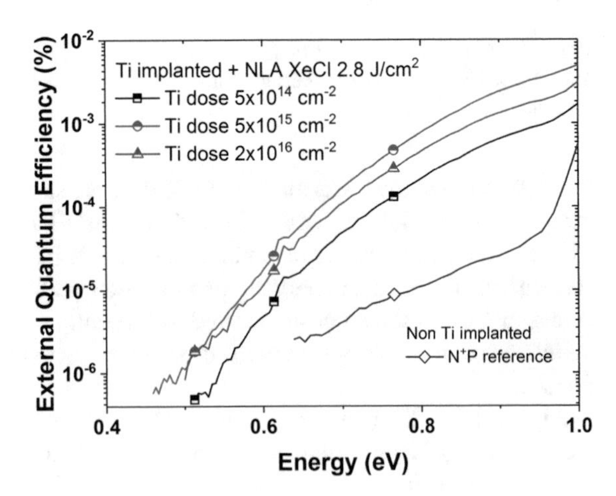

Fig. 4.34 External Quantum Efficiency, in the sub-bandgap region, as a function of the photon energy for Ti implanted samples at different doses, laser annealed at a fixed fluence of 2.8 J/cm^2. The reference N$^+$P diode is shown for reference

the sample with the middle Ti dose exhibit the highest EQE values, followed by the highest dose and then the lower Ti dose.

The role of Ti dose in the photogeneration of carriers is less accused than we could originally expect given the absortance results (Sect. 4.3.3) in the sub-bandgap region. According to the optical properties, the samples with the lowest Ti dose did not show any appreciable sub-bandgap absorption above the noise level of the measurement (set to approximately 0.8%), while the absortance of samples implanted with the highest Ti dose exhibited values in the order of 20% at 0.8 eV. Such difference in the photon optical absorption between different doses should lead to substantial differences in the sub-bandgap EQE. However, as stated before, the EQE measurements results from a compromise between the amount of generated carriers and the mean carrier lifetime, which directly depends on the crystal quality, among other parameters. The experimental data point to an optimal Ti dose which increases both the EQE for photon energies higher and lower than the bandgap. From the measurements shown in this sub-section, doses around 5×10^{15} cm^{-2} would provide the best results in the whole range, from UV to NIR. We suspect it could be due to a compromise between the mean carrier lifetime (severely influenced by the crystal quality of the TIL) and the optical absorption of photons.

Finally, we study the effect of the laser fluence on samples Ti implanted at a total dose of 5×10^{15} cm^{-2} in the sub-bandgap region.

As we did before, we separate the six Ti implanted samples with a total dose of 5×10^{15} cm^{-2} in the low and high energy regions. In none of the regions there seems to be a straight correlation between the sub-bandgap EQE and the laser fluence. In the low energy region (Fig. 4.35 left), the best device (taking the EQE value at 0.8 eV as reference) is the one annealed at 1.6 J/cm^2, while in the high energy region (Fig. 4.35 right) the best sample was annealed at 2.8 J/cm^2 of laser fluence. In general terms,

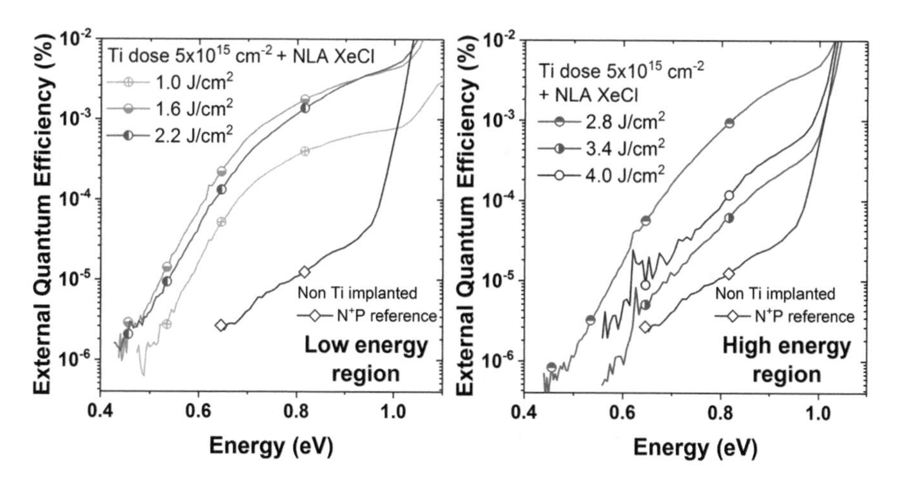

Fig. 4.35 External Quantum Efficiency, for sub-bandgap photons, as a function of the photon energy for Ti implanted at a dose of 5×10^{15} cm^{-2}, laser annealed at different laser fluences. The reference N^+P diode is shown for reference

samples within the low energy region showed higher sub-bandgap EQE values than samples belonging to the high energy region. In any case, all Ti implanted showed better EQE values than the non Ti implanted reference.

We propose three different mechanisms to explain the influence of the laser process in the sub-bandgap EQE. First, we propose the role of surface defects in the sub-bandgap photoresponse, mechanism that would be dominant in the reference sample, that was not Ti implanted. The contribution from surface defects is expected to be approximately the same in all the Ti implanted samples, as none of them is effectively passivated on neither surface. Several works studied the influence of surface defects in the sub-bandgap absorption, especially the work by Chiarotti et al. [40] where they proposed the existence of two different intermediate bands within the silicon bandgap related to surface defects. Secondly, we propose the role of bulk defects, coming from non-annealed defects after the NLA process, originated during the ion implantation process; the defects could generate deep levels into the Si bandgap [41, 42]. If the density of defects is high enough there could be an effective transfer of carriers from the valence band to the conduction band by means of tunnelling mechanisms. In cross-section HRTEM micrographs we saw that the TIL was partially nanocrystalline, partially amorphous in the sample laser annealed at 1.6 J/cm^2 (see Fig. 4.5). Raman measurements also pointed out to a higher defect density in samples laser annealed at fluences lower or equal than 2.2 J/cm^2 (see Fig. 4.3). Besides, the tunnel mechanism is further evidenced in the IV curves of those samples, where we obtained diode ideality factors higher than 2 (see Fig. 4.31), usually associated to this kind of tunnelling conduction mechanisms [37]. The sub-bandgap absorption contribution coming from bulk defects would be dominant on non-monocrystalline samples, i.e., the samples belonging to the low energy region (Fig. 4.35 left). Finally, the third contribution would account for the sub-bandgap carrier generation of Ti atoms, in concentrations high enough to overcome the IB formation limit, that is, the Ti supersaturated layer. As pointed out in our previous work by Olea et al. [43], the contribution from bulk defects to the sub-bandgap absorption in samples where the resultant TIL is monocrystalline (the high energy region) is of second order as compared to the contribution coming from Ti atoms.

It is possible to estimate the energy of the optical transition involved in the carrier photogeneration process by performing a Tauc plot of the EQE measurements [44]. In our case, we obtained the best fitting quality parameters using an indirect allowed transition [44]. We observed two different regimes that could be fitted, one close to the value of the bandgap of silicon, and a second one in the sub-bandgap region.

We fitted the experimental EQE data to a linear trend in the Tauc plot between 1.02 and 1.15 eV for the band-to-band transition, of all the Ti implanted samples implanted at different doses, laser annealed at different fluences (Fig. 4.36 left). We also performed the same analysis to the N$^+$P reference diode, where we obtained a value of 1.10 eV, compatible with the value obtained by fitting the absortance data, set to 1.09 eV, and very close to the referenced value of 1.12 eV. The energy of the band-to-band transition for Ti implanted samples is approximately constant, in the order of 1.07–1.08 eV for most samples, except for the sample with the lower Ti dose, annealed at a laser fluence of 1.6 J/cm^2, which showed a considerably lower value

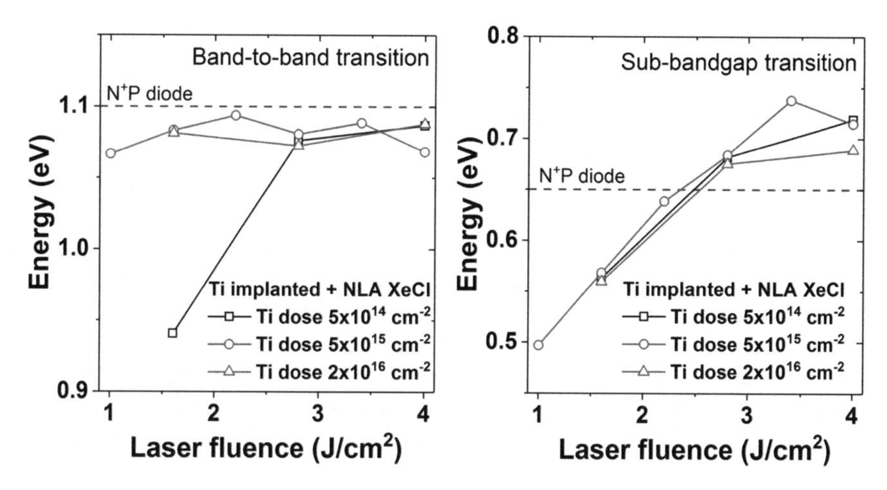

Fig. 4.36 Energy of the transitions obtained by applying a Tauc plot in the band-to-band region, between 1.02 and 1.15 eV (left figure) and in the sub-bandgap region, between 0.5 and 0.9 eV (right figure), as a function of the laser fluence for different Ti doses. The N^+P reference diode is included as a horizontal dashed purple line

of 0.94 eV, all of them showing lower values than the N^+P reference. Examining Fig. 4.36, we arrive to two main conclusions: the dose does not affect substantially the transition energy but the laser fluence does: the energy increases along with the laser fluence with almost the same trend for the three Ti doses. The energy varies from 0.49 eV for the lowest laser fluence up to 0.75 eV at high laser fluences.

Further research must be done in order to possibly identify the mechanisms behind the shift of the transition energy with increasing laser fluence. From the structural characterisation of the material we relate low transition energies (seen at low laser fluences) to poor crystal quality, fact that is supported in the stabilisation of the sub-bandgap transition energy in the high energy region, around a value close to 0.71 eV, where the TIL is monocrystalline.

Given the EQE measurements in the NIR range, all the Ti implanted samples exhibited sub-bandgap photoresponse at room temperature down to 0.45 eV (2.75 μm). The dose played a secondary role as compared to the effect of the laser fluence in the EQE. The best results were obtained for the sample Ti implanted at a dose of 5×10^{15} cm^{-2}, laser annealed at a fluence of 2.8 J/cm^2. We propose for future sample fabrication, using the XeCl laser, to narrow down the Ti implanted doses to a range between 10^{15} to 10^{16} cm^{-2}, with laser fluences in the order of 2.2–3.4 J/cm^2 to provide the best NIR EQE results.

4.6 Conclusions

Along this chapter we fabricated, measured and analysed samples Ti implanted in our facilities and laser annealed using a new UV excimer laser in our research group, under a collaboration with SCREEN-LASSE. The new laser uses XeCl to emit in the 308 nm line, featuring a longer pulse of 150 ns.

The material characterisation resulted in TIL with enough Ti concentration to supersaturate Si substrates, as measured by SIMS. We checked the crystal quality of the Ti implanted layers using cross-section HRTEM and Raman spectroscopy techniques. With the aid of these techniques we found two different melting regimes, depending on the laser fluence: the low energy regime, for laser fluences lower or equal than 2.2 J/cm^2, and the high energy regime, for fluence equal or higher than 2.8 J/cm^2. In the former, the energy absorbed by the sample is not enough to melt past the amorphous-crystalline interface, resulting in polycrystalline or amorphous layers after the solidification process. In the latter regime, the energy of the laser is high enough to melt the crystalline substrate. The recrystallization process takes a crystalline seed, resulting in monocrystalline regrown layers. These two regimes are compatible with the explosive recrystallization phenomenon, widely studied in the literature.

Later, we measured sub-bandgap absorption on the Ti implanted samples by means of T-R measurements. The sub-bandgap absorption depended mainly on the Ti dose, being higher with increasing Ti doses. Samples within the low energy regime exhibited interference patterns in the reflectance curves, which we associated to the presence of a highly defective layer close to the surface. In the high energy regime, the interferences almost disappeared, fact that could be used to estimate the onset of monocrystalline recrystallization using solely optical measurements. By applying a four-layer model, we could estimate the absorption coefficient around 10^3–10^4 cm^{-1}, results which are consistent with the previously published results in our research group.

Sheet conductance and Hall effect measurements were used to determine the electrical properties of the TIL, on n-type and p-type Si substrates. We observed that there was a transversal conductance that was dependent on the temperature. At high temperatures, the TIL and the substrate conducted in parallel. For temperatures lower than a certain threshold, the substrate decoupled from the TIL, visible as drastic change in the electrical properties of the samples. The conduction in the low temperature regime was dominated by the conduction inside the TIL. However, we observed several differences with respect to what was previously measured in our research group: there was another decoupling mechanism in some samples, which we attributed to the conduction through the Ti implanted but not laser annealed layer, within the TIL. Finally, we obtained that the TIL behaved as a metallic-like layer, with low Hall mobilities and high sheet concentrations.

The characterisation of the supersaturated material continued with the fabrication of transversal photodiodes on p-type substrates, the first prototypes used to demonstrate the viability of the technology, which were biased to obtain the current-voltage characteristics. We observed a rectification between the TIL and the substrate, suggesting that the TIL could behave as an n-type layer, similarly to what was obtained using four-probe measurements. The low diode ideality factors, around 1.5, obtained when applying the real diode model suggested that the conduction process could be driven by SRH recombination in the space charge region and in the neutral regions, if the laser fluence was high enough to produce a monocrystalline layer.

Finally, we observed lower EQE values in the UV and visible range in the Ti implanted samples when compared to the reference sample, in accordance to what was measured by our research group in the past. All the Ti implanted samples showed improved sub-bandgap photoresponse than the reference, the main goal of this thesis. We determined that Ti doses between 10^{15} and 10^{16} cm^{-2} and laser fluences in the range of 2.8–3.4 J/cm^2 could rend the best EQE results in the NIR range, while guaranteeing good electrical properties and good crystalline quality. These results are of extreme importance for the future design of the devices fabricated with collaboration of STMicroelectronics and CEA-LETI.

References

1. Deunamuno S, Fogarassy E (1989) A thermal description of the melting of c-silicon and a-silicon under pulsed excimer lasers. Appl Surf Sci 36:1–11
2. Ishihara R, Yeh WC, Hattori T, Matsumura M (1995) Effects of light-pulse duration on excimer-laser crystallization characteristics of silicon thin-films. Jpn J Appl Phys 1(34):1759–1764
3. Olea J, del Prado A, Pastor D, Martil I, Gonzalez-Diaz G (2011) Sub-bandgap absorption in Ti implanted Si over the Mott limit. J Appl Phys 109
4. Dawber PG, Elliott RJ (1963) Theory of optical absorption by vibrations of defects in silicon. Proc Phys Soc 81:453–460
5. Echlin P (2009) Handbook of sample preparation for scanning electron microscopy and x-ray microanalysis. Springer, USA, p 332
6. Poppendieck TD, Ngoc TC, Webb MB (1978) An electron diffraction study of the structure of silicon (100). Surf Sci 75:287–315
7. Williams DB, Carter CB (1996) Transmission electron microscopy. Springer
8. Moretti G (1998) Auger parameter and Wagner plot in the characterization of chemical states by X-ray photoelectron spectroscopy: a review. J Electron Spectrosc Relat Phenom 95:95–144
9. Henderson R (1972) Silicon cleaning with hydrogen peroxide solutions: a high energy electron diffraction and Auger electron spectroscopy study. J Electrochem Soc 119:772–775
10. Donovan EP, Spaepen F, Turnbull D, Poate JM, Jacobson DC (1983) Heat of crystallization and melting-point of amorphous-silicon. Appl Phys Lett 42:698–700
11. Green MA (2008) Self-consistent optical parameters of intrinsic silicon at 300 K including temperature coefficients. Sol Energ Mat Sol C 92:1305–1310
12. Hishikawa Y et al (1991) Interference-free determination of the optical-absorption coefficient and the optical gap of amorphous-silicon thin-films. Jpn J Appl Phys 1(30):1008–1014
13. Moon S, Minghong L, Grigoropoulos CP (2002) Heat transfer and phase transformations in laser annealing of thin Si films. J Heat Transf 124:12
14. Aydinly A, G. Y, Topaçli C (1988) Simulation of explosive crystallisation in pulsed laser irradiated a-Si. J Appl Phys 22:6
15. Gotz G (1986) Explosive crystallization processes in silicon. Appl Phys a-Mater 40:29–36
16. Thompson MO et al (1984) Melting temperature and explosive crystallization of amorphous-silicon during pulsed laser irradiation. Phys Rev Lett 52:2360–2363
17. Thompson MO et al (1983) Silicon melt, regrowth, and amorphization velocities during pulsed laser irradiation. Phys Rev Lett 50:896–899

18. Miki T, Morita K, Sano N (1997) Thermodynamic properties of titanium and iron in molten silicon. Metall Mater Trans B 28:861–867
19. Lowndes DH et al (1985) Direct imaging of "explosively" propagating buried molten layers in amorphous silicon using optical, Tem and Ion backscattering measurements. MRS Proc 51:131
20. Baeri P, Rimini E (1996) Laser annealing of silicon. Mater Chem Phys 46:169–177
21. Bonafos C, Mathiot D, Claverie A (1998) Ostwald ripening of end-of-range defects in silicon. J Appl Phys 83:3008
22. Smit C et al (2003) Determining the material structure of microcrystalline silicon from Raman spectra. J Appl Phys 94:3582–3588
23. Maley N (1992) Critical investigation of the infrared-transmission-data analysis of hydrogenated amorphous-silicon alloys. Phys Rev B 46:2078–2085
24. Edwards DF, Ochoa E (1980) Infrared refractive index of silicon. Appl Opt 19:4130–4131
25. Jellison GE Jr, Lowndes DH (1987) Measurements of the optical properties of liquid silicon and germanium using nanosecond time-resolved ellipsometry. Appl Phys Lett 51:352–354
26. Olea J (2009) Procesos de implantación iónica para semiconductores de banda intermedia. Thesis dissertation
27. Garcia-Hemme E (2015) Respuesta infrarroja en silicio mediante implantación iónica de metales de transición. Thesis dissertation
28. Olea J et al (2012) Low temperature intermediate band metallic behavior in Ti implanted Si. Thin Solid Films 520:6614–6618
29. Castan H et al (2012) Electrical properties of intermediate band (IB) silicon solar cells obtained by titanium ion implantation. AIP Conf Proc 1496:189–192
30. Pastor D et al (2012) Insulator to metallic transition due to intermediate band formation in Ti-implanted silicon. Sol Energ Mat Sol C 104:159–164
31. Pastor D et al (2013) Electrical decoupling effect on intermediate band Ti-implanted silicon layers. J Phys D Appl Phys 46
32. Garcia-Hemme E et al (2013) Electrical properties of silicon supersaturated with titanium or vanadium for intermediate band material. Proceedings of the 2013 Spanish Conference on Electron Devices (CDE 2013), 377–380
33. Olea J et al (2011) Two-layer Hall effect model for intermediate band Ti-implanted silicon. J Appl Phys 109
34. Garcia-Hemme E et al (2015) Meyer Neldel rule application to silicon supersaturated with transition metals. J Phys D Appl Phys 48
35. Olea J, Pastor D, Toledano-Luque M, Martil I, Gonzalez-Diaz G (2011) Depth profile study of Ti implanted Si at very high doses. J Appl Phys 110
36. How H, Weidong T, Vittoria C (1997) AC-Hall effect in multilayered semiconductors. J Light Technol 15:1006–1011
37. Neamen DA (1997) Semiconductor physics and devices, vol 3, McGraw-Hill, New York
38. Gambino JP, Colgan EG (1998) Silicides and ohmic contacts. Mater Chem Phys 52:99–146
39. Olea J et al (2016) Room temperature photo-response of titanium supersaturated silicon at energies over the bandgap. J Phys D Appl Phys 49
40. Chiarotti G, Nannarone S, Pastore R, Chiaradia P (1971) Optical absorption of surface states in ultrahigh vacuum cleaved (111) surfaces of Ge and Si. Phys Rev B-Solid St 4:3398
41. Casalino M, Coppola G, Iodice et al (2010) Near-infrared sub-bandgap all-silicon photodetectors: state of the art and perspectives. Sensors 10:10571–10600
42. Fan H, Ramdas A (1959) Infrared absorption and photoconductivity in irradiated silicon. J Appl Phys 30:1127–1134
43. Olea J et al (2013) Ruling out the impact of defects on the below band gap photoconductivity of Ti supersaturated Si. J Appl Phys 114
44. Viezbicke BD, Patel S, Davis BE, Birnie III DP (2015) Evaluation of the Tauc method for optical absorption edge determination: ZnO thin films as a model system. physica status solidi (b) 252:1700–1710

Chapter 5
Results: Integrating the Supersaturated Material in a CMOS Pixel Matrix

5.1 Introduction

The information contained in this chapter is the result of the collaboration between the UCM and STMicroelectronics, through an internship financed by the Spanish Ministry of Science and Universities, under grant no. EEBB-I-17-12315. The internship took place between September 1st of 2017 and March 4th of 2018.

The aim of the internship was to study de integrability of the Ti supersaturated Si layer within a pre-commercial pixel matrix based on CMOS Image Sensor (CIS) technology, developed by STMicroelectronics in Crolles (France). The market is currently demanding NIR and SWIR detectors with reduced cost and high scalability. As described in the previous chapters, it is possible to fabricate Si with improved responsivity in the IR, through Ti supersaturation of Si wafers. The fact that a company as important as STMicroelectronics has shown interest in the technology developed in our research group demonstrates that the concept of supersaturated materials could be ground-breaking in the world of IR sensing. Most of the details concerning the integration and structure of the final devices lies within a Non-Disclosure Agreement (NDA). Therefore, this chapter will omit most of the information related to pixel structure, specific fabrication steps or any other aspect subscribed under said NDA document. The information of this chapter will be focused in the material characterisation and a brief description of some results derived from the internship.

The collaboration project is divided into four main blocks: first, the study and design of the integration protocol of the Ti supersaturated layer into the existing pixel structure. Each individual extra step must be thoroughly studied, revised and validated by different experts inside the fabrication chain prior to the execution. Second, the execution of the designed workflow. Third, the qualification and characterisation of the extra fabrication steps introduced to integrate the supersaturated material, and fourth, the characterisation of the manufactured material and devices.

D. Montero Álvarez, *Near Infrared Detectors Based on Silicon Supersaturated with Transition Metals*, Springer Theses,
https://doi.org/10.1007/978-3-030-63826-9_5

The development of a research project in a multinational company, with several thousands of employees in the same worksite, is radically different from what is done in a research group of a university. Every single design, manufacturing or characterisation step is carefully optimised, so that only trained personnel with special machinery is capable to perform a certain task. That said, my work within the project was to design and successfully coordinate and transfer the information to any link of the fabrication and characterisation chain: from the first concepts to the small details concerning handling, dicing or labelling each wafer or device. I was in charge of the data analysis and part of the decision making. Most of the characterisation (in particular the Wafer Level Characterisation measurements) were performed by automated instruments. All the die-per-die characterisation was performed manually by myself, as well as part of the material characterisation that was done at UCM after the internship was over (sheet resistance, AFM and TEM mostly).

5.2 Pixel Structure

The device architecture used to integrate the Ti supersaturated layer is a prototype vehicle test chip, the pre-commercial stage of development of a CMOS Image Sensor. The structure of this type of devices is close to the final version, but it still needs improvements and fine tuning of some fabrication steps and/or parts of the device structure. Generally, the pre-commercial devices are simpler, acting as vehicles to demonstrate the viability of the technology.

When it comes to pixel design, there are two main types of pixels, according to the direction of the incident illumination, the FrontSide Illumination (FSI) pixel, and the BackSide Illumination (BSI) pixel. FSI pixels are illuminated from the same surface in which the transistors and ROIC are located, while BSI pixels are illuminated from the opposite surface with respect to the transistors and ROIC. FSI pixels are easier to manufacture (they require less fabrication steps), but the transistors and ROIC produce a shadow over the active area of the pixel, which in turn increases the area of each pixel. BSI pixels were developed to increase the effective pixel area [1]. As the size of the pixel was being diminished to fulfil the demands of higher resolution in smaller chips, the BSI configuration offered the highest ratio of useful area over the total pixel area: the transistors and ROIC, by being on the opposite surface of the pixel, did not interfere with the incoming rays (Fig. 5.1).

Two possible pixel structures were proposed by STMicroelectronics for integration with the Ti supersaturated material, both pixels using the Back Surface Illumination (BSI) configuration: an electron collection pixel, based on Vertically-Pinned Photodiodes (V-PPD) [2] and a hole collection pixel, based on a vertical isolated gate transistor [3]. Based on the n-type like conduction mechanisms found on Ti supersaturated layers over p-type substrates, at room temperature (Sect. 4.5.1), we decided to choose the hole collection pixel due to the expected carrier migration direction inside the pixel structure.

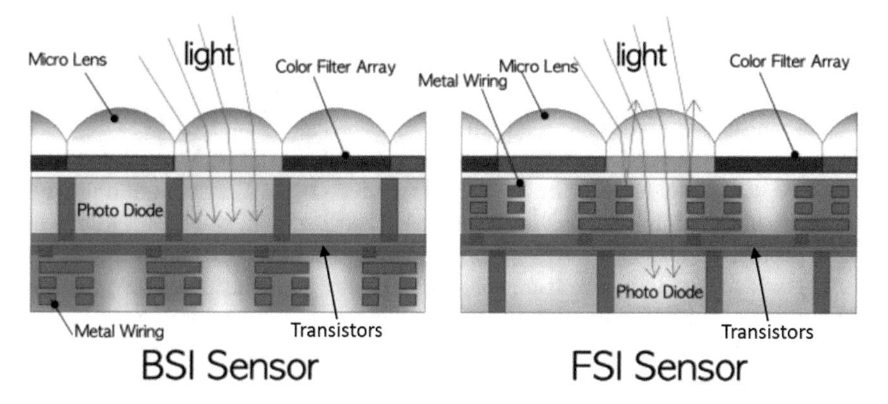

Fig. 5.1 Front-Side Illuminated pixel (left) and Back-Side Illuminated pixel cross-section schematics. Transistor area is indicated by a red rectangle

The project inside the company had assigned a total of eleven 300 mm wafers already processed with hole collection pixels (referred to as patterned wafers in the text). Additionally, another lot of ten 300 mm blank wafers was included in the project, aiming to be used as testing wafers on the various development stages of the project. These wafers are referred to as monitor wafers in the text. The base substrate used in this chapter is the one coded as P3 in Table 2.1.

Each portion of semiconductor containing a functional pixel matrix, along with its ROIC is usually labelled as a "die". Usually, each patterned wafer may contain from several tens to several thousand dies that followed the same fabrication process. The architecture used on each hole collection pixel matrix holds up to 10^6 pixels (1 megapixel), in a squared-shaped sensor area of 1000 pixels per side, which is sub-divided into ten different groups of pixels. Each group has different pixel or ROIC versions, which are part of the NDA, but they are usually related to different transistor versions, different doping levels or different features of the pixel with varying sizes. Pre-commercial CMOS image sensors usually benefit from their large number of pixels to sub-divide them into different versions, aiming to speed up the development process by sweeping several parameters in the same run in search for the best performance. Therefore, the amount of measured pixels per wafer may be in the order of 10^7–10^9, which requires advanced computing and statistical techniques to analyse the results.

5.3 Integrating the Ti Supersaturated Layer

After deciding the pixel structure, it is time to study the different possibilities of integration of the TIL in the fabrication route of the pixel matrix. Given that industrial processes are quite complex and have been carefully optimised, modifying or introducing additional fabrication steps is not straight-forward. The operators must have the certainty that it won't affect the rest of the fabrication processes inside the fab.

The fabrication processes in a CMOS route are mainly divided in two groups: Front-End Of Line (FEOL), which comprises from the blank wafer up to the fabrication of the transistors and the Back-End Of Line (BEOL), where the transistors are interconnected, the passivation layers are deposited and, in general terms, the wafer processing is finished. After, there are further back-end operations like dicing, packaging and labelling, which are done in other facilities, where the individual dies are cut and wire bonded to a chip carrier so that they can be properly used in their final application. The main concern when introducing Ti-rich layers inside a fabrication line (in the BEOL) is the cross-contamination of samples of other lots coming from our project. Therefore, a lot of resources were put to investigate the possible consequences and how to avoid cross-contamination coming from Ti atoms. Several monitor wafers were used for that matter.

The use of BSI pixels was advantageous when integrating the TIL: there are no transistor nor other important structures on the backside that could be affected by the ion implantation nor the NLA process. The wafers are received from the FEOL with the transistors fabricated (including the necessary thermal processes to activate the dopants of transistors + ROIC) in the frontside of wafer 1. In order to eliminate cross-talk, there are trenches between pixels that isolates them, of the same height as the pixel, around 4 μm. The back surface of the pixel must be thinned down to the said pixel thickness value, which requires etching 745 μm from the original wafer. To do so without compromising the integrity of the structure, a second wafer must me bonded on the frontside of the pixel structure (where transistors and ROIC are located). After the Chemical-Mechanical Polishing (CMP) process the backside of the pixel is exposed. It is in this surface where we implant Ti, which is later NLA processed (we refer to this surface as the frontside of wafer 1 + wafer 2). Since the heat only penetrates several hundreds of nanometres, the transistors + ROIC are not affected by the laser process. After the NLA process, the next steps are the deposition of the ARC layer, a last metallisation process, a FGA and the deposition of colour filters and microlenses. No important annealing processes (higher than 400 °C) need to be done to finish the pixel structure that could severely compromise the optoelectronic properties of the TIL. Other studies with supersaturated Si discourage the use of conventional thermal processes (as RTA or FGA) to avoid the deactivation of the supersaturated layer [4, 5]. Therefore, the best option for integration was to implant Ti atoms in the exposed rear surface of the pixel prior the deposition of the ARC layer (Fig. 5.2).

After the Ti implantation process, which was done at Ion Beam Services in Peynier (France) because of their compatibility with 300 mm wafers, it was necessary to recover the crystal quality using a NLA process. After the signing of the collaboration and the NDA document with the company, part of the work of this thesis was to find a suitable laser certified for 300 mm technology, in order to develop the Ti supersaturated material with the collaboration of STMicroelectronics. We found the XeCl NLA equipment described in Sect. 2.1.4. The purpose of the Chapter 4 was

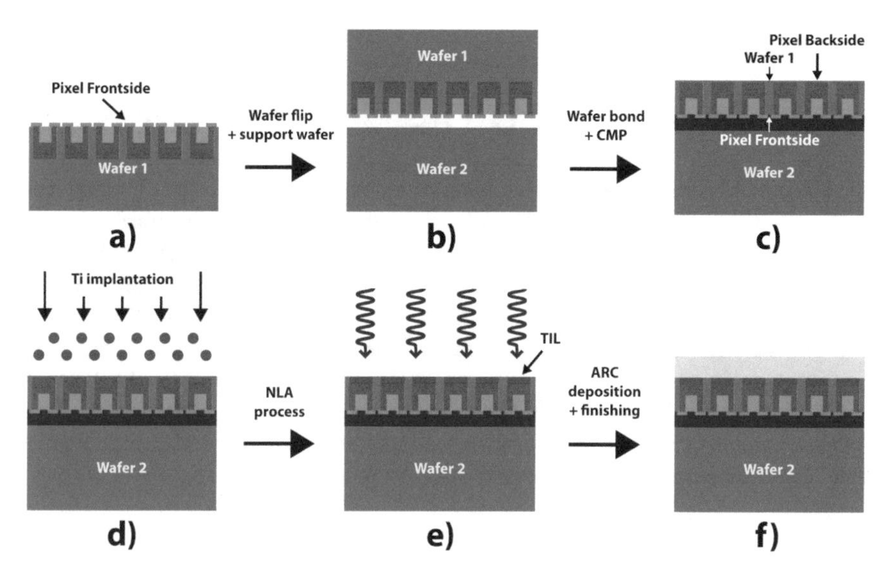

Fig. 5.2 Cross-section schematic of the fabrication process of BSI pixels, along with the Ti integration. **a** Starting wafer as received from the FEOL, **b** Wafer alignment for mechanical support, **c** Wafer bonding + Chemical-Mechanical Polishing of back surface of wafer 1, **d** Ti implantation on the backside of the pixel (frontside of wafer 2), **e** Laser annealing of the Ti implanted layer and **f** ARC deposition + metallisation + finishing steps

to fully characterise the properties of the Ti supersaturated layer with the new laser. Thus, for samples described in this chapter, we used the same NLA equipment, the XeCl laser, which was relocated at CEA-LETI facilities in Presqu'île Grenoble (France), less than 30 km away from STMicroelectronics facilities. Before and after both the Ti implantation and the NLA processes, the Ti contamination was checked to ensure the operation of the manufacturing machinery within safety limits. All the samples belonging to the project (11 patterned wafers plus 10 monitor wafers) were Ti implanted on the front-side of the wafer. All patterned wafers and some monitor wafers were laser annealed.

The wafers then returned to the normal BEOL processes like non Ti implanted wafers. For this project, the last steps were the ARC deposition of an Oxide-Nitride-Oxide (ONO) triple layer, the etching of some pits and metallic deposition to contact the transistors on the other side of the pixel and finally a FGA of 380 °C during 2 h, necessary to passivate the surface and to activate the metallization. This last step was unavoidable, and its possible influence on the performance of the TIL is yet undetermined. No colour filters nor microlenses were deposited in the samples studied in this chapter.

5.4 Material Fabrication

Now that the integration plan has been defined, we focus in the material fabrication process. First, it is necessary to choose the Ti ion implantation parameters. As we have seen in the previous chapters, doses equal or higher than 5×10^{15} cm^{-2} could lead to the apparition of CBD structures. This might lead to an increase of pixel response variability that might affect Photon Response Non-Uniformity (PRNU) and Dark Signal Non-Uniformity (DSNU). Besides, although crystal quality was shown to be good for these Ti doses, it is expected to find more defects when increasing the Ti dose, as seen by Raman spectroscopy on samples in the previous chapter (Fig. 4.3). For lower doses, close to 5×10^{14} cm^{-2}, the absorption coefficient (as displayed in Fig. 4.9) is too low. Besides, the IB layer thickness would be very thin so the expected enhancement in QE could be lower than desired. Given that, and taking into account that 11 wafers were available, the next doses were chosen:

$$\text{Ti dose}\begin{cases} 1 \times 10^{15}\text{cm}^{-2} \\ 2 \times 10^{15}\text{cm}^{-2} \\ 3 \times 10^{15}\text{cm}^{-2} \\ 5 \times 10^{15}\text{cm}^{-2} \end{cases}$$

Once we have decided the number of different Ti doses, we defined the split plan, assigning to each Ti dose a certain number of wafers. We decided to use 3 wafers per dose, except for the dose of 10^{15} cm^{-2} in which we used only 2. With respect to monitor wafers, we used 2 wafers per dose except for the highest dose, 5×10^{15} cm^{-2} in which four were made. It was expected that, if any contamination were found, the highest dose would be the worst case scenario, so it was used for testing in all equipment before the rest of the doses. If no contamination were to be found on wafers with the highest Ti dose (no Ti contamination on neither the wafers themselves or the inspection tools) then the rest of wafers were expected to be non-contaminant.

In the previous chapter we identified a possible problem: the ion implantation energy was too high, as the laser could not melt the whole Ti implanted layer (as evidenced by SIMS measurements Fig. 4.2). For the pixels, we wanted to be sure to melt all Ti atoms to avoid the formation of defects. To do so, we lowered the ion implantation energy to levels in which we were sure the laser was capable to melt. It was extremely important to melt up to the deepest Ti atom, to minimize the concentration of deep centers that could drastically reduce minority carrier lifetime and increase the dark current of the pixels. We estimated the melted depth of the NLA process to be around 260 nm for crystalline Si at 3.6 J/cm^2 of laser fluence, using the experimental data of K. Huet et al., where they used the same LT3100 laser equipment used in this thesis [6]. Therefore, we chose an ion implantation energy low enough to assure that most Ti atoms were implanted at depths lower than 260 nm. After the simulation of different Ti implantation profiles using SRIM, we chose an ion implantation energy of 10 keV. As usual, we chose the tilt implantation angle at 7°. The implanted area was limited by the configuration of the ion implanter, set to a circular area of around 180 mm diameter, almost centered with respect to the wafer. This is a radical difference with respect to the previous chapter, were we used ion implantation energies of 150 keV.

The ion implantation of only a fraction of the wafer surface implied that less devices were Ti implanted than originally expected. The bright side was the possibility to study the effect of the laser process on non Ti implanted devices in the same wafer. This way, on each wafer, we could have reference dies (non-implanted, non-annealed), non-implanted dies but laser annealed, and Ti implanted and laser annealed dies. Thus, it would be possible to separate the effect of the laser process and the Ti implantation on the performance of the device.

The next step in the fabrication route is the NLA process. Prior to laser annealing the patterned wafers, we used the monitor wafers to test the laser recipes, in particular the effect of the laser fluence. The configuration of the laser at the moment of annealing had the focusing system set up so that a square-shaped area of 15×15 mm was irradiated, with a maximum laser fluence of 3.6 J/cm^2. In the first run, we laser annealed the monitor wafers plus one non Ti implanted wafer as reference. Each wafer contained 76 different exposed areas (also referred to as dies), spaced 3 mm, divided in two groups of 38. The first group received only one shot at an increasing laser fluence, starting from 0.80 up to 3.60 J/cm^2, with steps of 75 mJ/cm^2. The second group was fabricated firing two laser shots on each die. The first shot followed the same sequence as in the first group, sweeping from 0.80 to 3.60 J/cm^2. The second laser shot was fired for all 38 dies of the second group at a fixed value of 2.4 J/cm^2, aiming to study the properties of the material after the second shot, which could help determining the recrystallization regimes of the Ti implanted samples. A brief analysis followed the NLA process of the monitor wafers, which was key to decide what laser fluences should be used in the patterned wafers. Among the available data coming from the monitor wafers, we analysed the Time-Resolved Reflectometry and the Haze Measurements, fast and in situ wafer-level measurements.

Before laser processing the patterned wafers, it was necessary to define the split plan of the laser process. To do so, we needed to find out first how many dies were implanted, since the Ti implanted area was smaller than the wafer diameter. Considering the floor plan of the architecture and the chosen interval of fluences, we designed the split plan.

Figure 5.3 shows the location and laser fluence used to anneal each die inside the wafer. The laser fluence map is thought in such way that most dies with same laser fluence are not located close to each other, in order to minimise the effect of possible heterogeneities that could affect to all the dies belonging to the same laser fluence. Lower laser fluences are indicated by cold colours, starting from dark blue, up to the highest laser fluence, indicated by warm colours. Most dies within the Ti implanted areas were annealed using laser fluences higher than 2.0 J/cm^2, a threshold value that was obtained after analysing the monitor wafers and that is also close to the value obtained in the previous chapter. As some dies may be partially implanted, high variability may be expected for dies in the limit of the implanted region. We list the amount of available dies as a function of the laser fluence in Table 5.1.

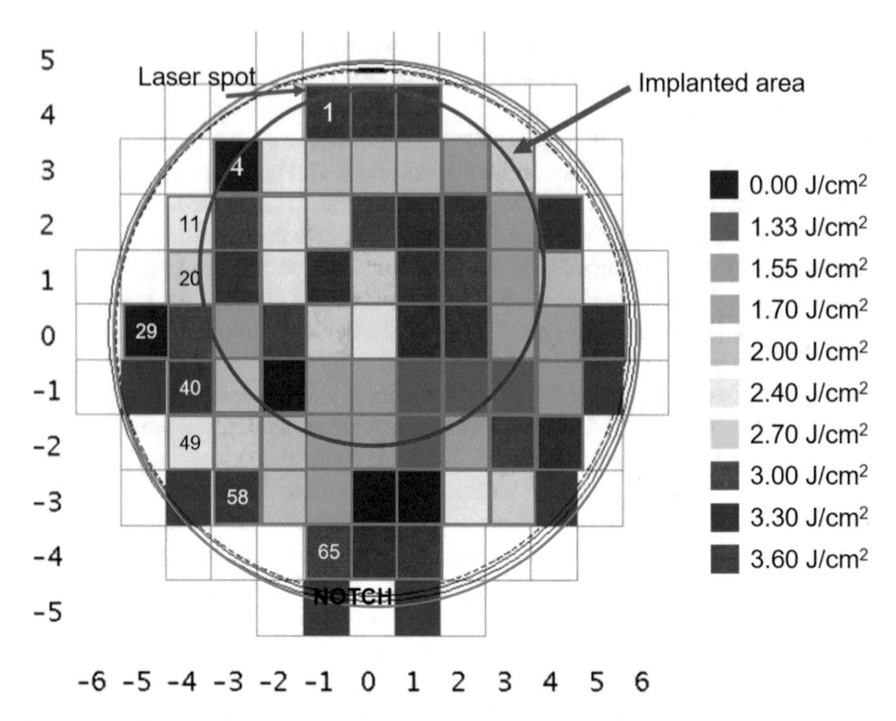

Fig. 5.3 Split plan of the NLA process on 300 mm patterned wafers. The red circle delimits the Ti implanted area. Each colour indicates a different laser process for each die. Laser fluence is increasing from dark blue to magenta

Table 5.1 Die laser process distribution: number of Ti implanted a non-implanted dies for each laser fluence

Laser fluence (J/cm^2)	Implanted dies	Non-implanted dies
No NLA process	1	4
1.33	2	2
1.55	4	3
1.70	4	3
2.00	3	4
2.40	4	3
2.70	4	3
3.00	4	3
3.30	3	4
3.60	7	3
Total	36	32

After the laser process, done at CEA-LETI facilities, the wafers were sent back to STMicroelectronics fab to continue the fabrication process. Only patterned wafers will continue inside the fab. Monitor wafers are stored in the cleanroom for further material characterisation. The next step was the deposition process of the triple ONO layer using Plasma Enhanced Chemical Vapour Deposition (PECVD) technique at 380 °C.

The main issue regarding the deposition steps is the metallic contamination to either the deposition chamber or other wafers inside the fab. Ti atoms could be sputtered to the chamber during deposition, or they could simply diffuse towards the surface due to the temperature increase. Several studies of Ti diffusion on Si samples were considered, although the majority of them dealt with bulk Ti diffusion [7] at temperatures ranging from 600 to 1200 °C [8]. Since it is generally assumed that diffusion phenomena in Si follow an Arrhenius law, the experimental data were fitted to an exponential decay and the diffusion value at 400 °C was extrapolated. The extrapolated values ranged between 5.74×10^{-15} cm^2/s and 1.81×10^{-19} cm^2/s, which, at this temperature, is placed between the diffusion coefficient of phosphorus and boron [9]. However, in another work [10] they studied Ti diffusion through the surface, which could be the closest experimental evidence to our concern of Ti contamination. Data are extracted from the last paper and represented in Fig. 5.4. From fitting coming from Fig. 5.4, the diffusion coefficient at 400 °C was extrapolated to 5.32×10^{-17} cm^2/s, value that lies in the middle of the range estimated a few lines back using other references. According to these estimations, Ti out-diffusion should be negligible during the deposition process. However, sputtering of Ti atoms during the first stages of the deposition process cannot be completely discarded.

The rest of fabrications steps did not substantially affect the Ti supersaturated layer, so they will be omitted here. As a final consideration of the material fabrication, the total thermal budget of the ONO deposition process, although at relatively low temperatures, could affect the IB material in an unpredictable way. We expect to fabricate a new set of samples with similar ion implantation parameters in order to

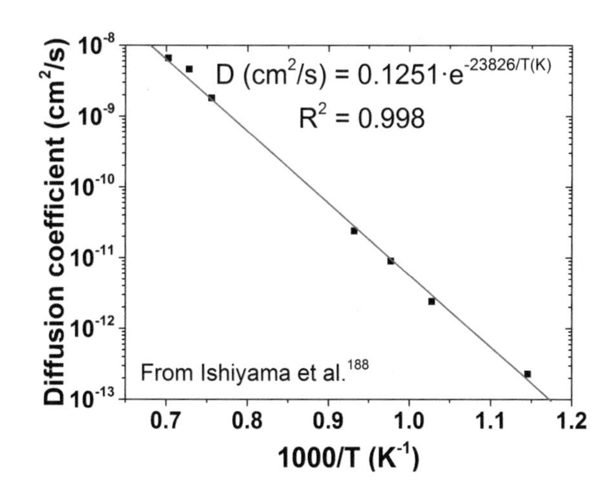

Fig. 5.4 Ti diffusion coefficient on Si at several temperatures. Fitting has been done to an Arrhenius law. Data extracted from Ishiyama et al. [10]

measure the optoelectronic properties before and after similar thermal processes to analyse the possible effect of the last FGA process.

5.5 Material Characterization

The previous section briefly described the different steps followed during the material fabrication. In this section, we analyse in depth the different characterisation techniques and the derived results from them to further understand the properties of the Ti supersaturated layer. The material characterisation will be focused on addressing the crystallinity of the TIL as a function of the laser fluence. In a photodiode, the best crystal quality is always preferred due to the associated lower dark current generation. As we have discussed before in Sect. 2.6.1, dark current is the main mechanism degrading the performance of a pixel, and also a source of inhomogeneities, both undesirable effects.

5.5.1 Time-Resolved Reflectometry

The first technique used to characterise the supersaturated layer was the TRR, a measurement that is triggered at the same time as when the NLA process is taking place on the sample. TRR measures the change in reflectance of the sample as a function of time. Comparing the different TRR curves on monitor wafers, with different laser fluences, it is possible to estimate the different melting regimes for each Ti dose. This relatively fast and non-destructive technique allowed us to choose the appropriate laser fluence range to be used in the patterned wafers, which were annealed shortly after the monitor lot.

For clarity, we show only the measurements done on one of the monitor wafers with a Ti dose of 3×10^{15} cm^{-2}. After the analysis of said Ti dose, we study the influence of the Ti dose in the recrystallization regimes. A similar behaviour is found for all Ti doses.

TRR curves, displayed in Fig. 5.5, exhibited up to four different behaviors. From 0.80 to 1.10 J/cm^2, TRR profiles did not display any significant feature. Voltage values measured by the reflectivity sensor remained almost flat around an average value of 0.32 V. We found a similar behavior in non-implanted Si samples annealed using the same fluence interval. We have labelled this region as the "sub-melt" regime.

Second regime, labelled as "First melt and solidification" appears when energy is increased beyond 1.10 J/cm^2. TRR curves display an asymmetric peak around 100 to 200 ns from trigger, which we have labelled as α in Fig. 5.5. The α peak appears at shorter times, displaying also higher reflectivity values, as the laser fluence increases. Up to 1.40 J/cm^2 this is the only peak observed. Non-implanted samples did not show any observable peak.

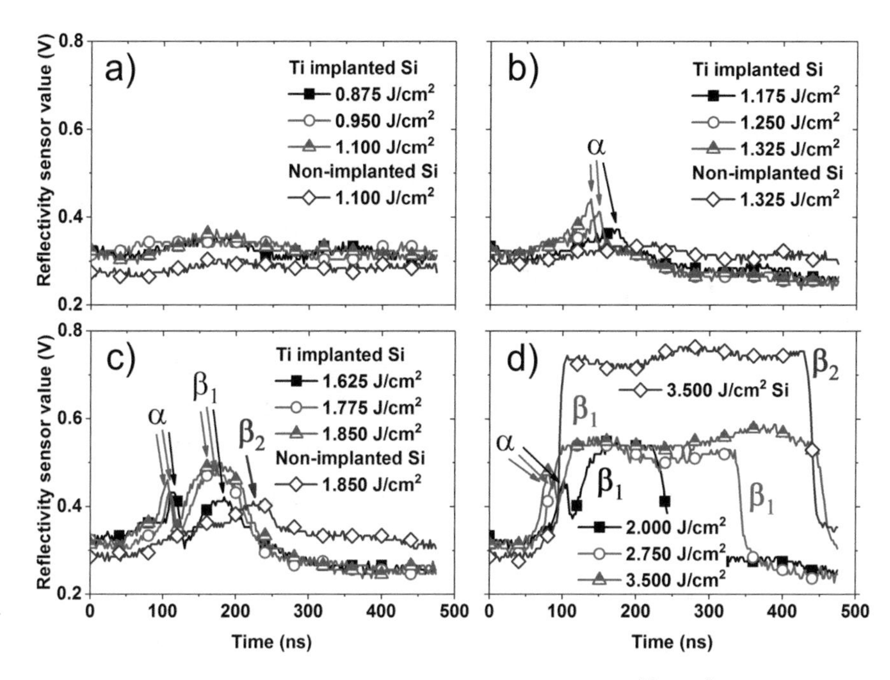

Fig. 5.5 TRR curves of Ti implanted samples at a dose of 3×10^{15} cm^{-2}, laser annealed at different laser fluences. They are divided into four main regions: sub-melt regime (**a**), first melt and solidification (**b**), second melt and crystallisation onset (**c**) and high energy regime (**d**). The TRR of a non Ti implanted sample has been included for comparison

Third regime, the "second melt and crystallization onset", starts at 1.40 J/cm^2 and extends up to 2.00 J/cm^2. Its main characteristic is the apparition of a second peak, after the α peak, for times around 170 ns from the origin. We have labelled the new peak as the β_1 peak, also indicated in Fig. 5.5. This peak appears to follow the same trend as the α peak: when we increase the laser fluence, the peak appears at shorter times, and the maximum reflectivity value at the peak is higher as well. Besides, peak width is increasing along with the laser fluence, due to a displacement of the rise time to shorter times, in addition to the fall time moving to longer times. In this regime, for laser fluences equal or higher than 1.85 J/cm^2, non-implanted dies showed a peak, labelled as the β_2 peak.

At the "high energy" regime, the β_1 peak enlarges its width, evolving into a plateau, instead of showing a unique maximum value. The plateau reflectivity value of Ti implanted samples saturates at around 0.5 V in all TRR curves measured from 2.00 J/cm^2 to 3.58 J/cm^2. Non-implanted c-Si exhibits the same behavior in this region, as the β_2 peak also evolved into a plateau, but with a saturation value much higher than β_1 peak, around 0.73 V, as measured by the sensor.

After analyzing the TRR curves, we observe that α and β_1 peaks exhibit different properties, in particular, the width and the maximum reflectance values. The width of the α peak remains approximately constant for all laser treatments, with values

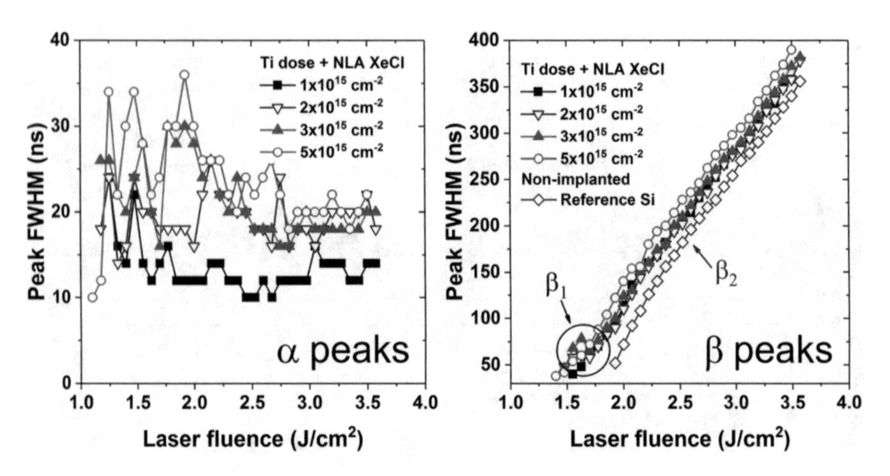

Fig. 5.6 Peak FWHM of α, β_1 and β_2 peaks/plateaus as a function of the laser fluence for different Ti doses and the non-implanted reference

between 35 and 10 ns (see Fig. 5.6), while the maximum reflectance value increases from 0.3 up to 0.45 V (see Fig. 5.5). In contrast, the width of the β_1 peak increases along with the laser fluence, evolving into a plateau for laser fluences higher than 1.93 J/cm^2, featuring a maximum width of 382 ns at the highest value. The reflectance values ranged from 0.3 to 0.5 V, higher values than in the case of the α peak, in the high energy regime. With respect to non-implanted samples, we performed a similar analysis as we did with Ti implanted samples. In comparison to α or β_1 peaks, β_2 peaks and plateaus appeared in the TRR curves at higher laser energies. Examining the maximum sensor reflectivity value at the peak, we found a continuous increase in reflectivity up to 2.00 J/cm^2, where it stabilizes around 0.75 V, a value substantially higher than the one found for α peaks or β_1 plateaus. Finally, peak duration was found to be strongly dependent on the laser energy density, as it was found for the β_1 peak of Ti implanted samples (Fig. 5.6).

Figure 5.6 left shows the effect of the laser fluence on the duration of the α peak for the four Ti doses contained in this chapter. The peak FWHM does not change substantially when increasing the laser fluence, being the curves quite noisy. There seems to be an influence of the Ti dose, although it is not that clear at laser fluences higher than 2.0 J/cm^2. There is a clear influence of the Ti dose on the peak FWHM curves for β_1 and β_2: at a given laser fluence, higher Ti doses led to thicker peaks. All the curves related to β_1 and β_2 peaks showed a linear-like trend with respect to the laser fluence. This result is particularly interesting, as in other references a quadratic dependence was found when laser annealing Si samples implanted with other chemical species, such as As [11], although a different laser was used for this study.

5.5.2 *Haze Measurements*

After the NLA process, we measured the Haze of the Ti implanted monitor wafers at the CEA-LETI facilities. These measurements were used to further check the different recrystallization regimes on monitor wafers, aiming to adequately select the appropriate laser fluence range for the patterned wafers. We obtained the Haze map for each Ti dose (Fig. 5.7).

The laser fluence is increasing when going to the right in the same row and when going to lower rows within the same column. The 76 laser shots are visible in the Haze map, from the two groups previously described of 38 laser shots each. In order to quantify the Haze value inside each exposed area, we took five points inside the die. We represent the average value and the standard deviation as the uncertainty:

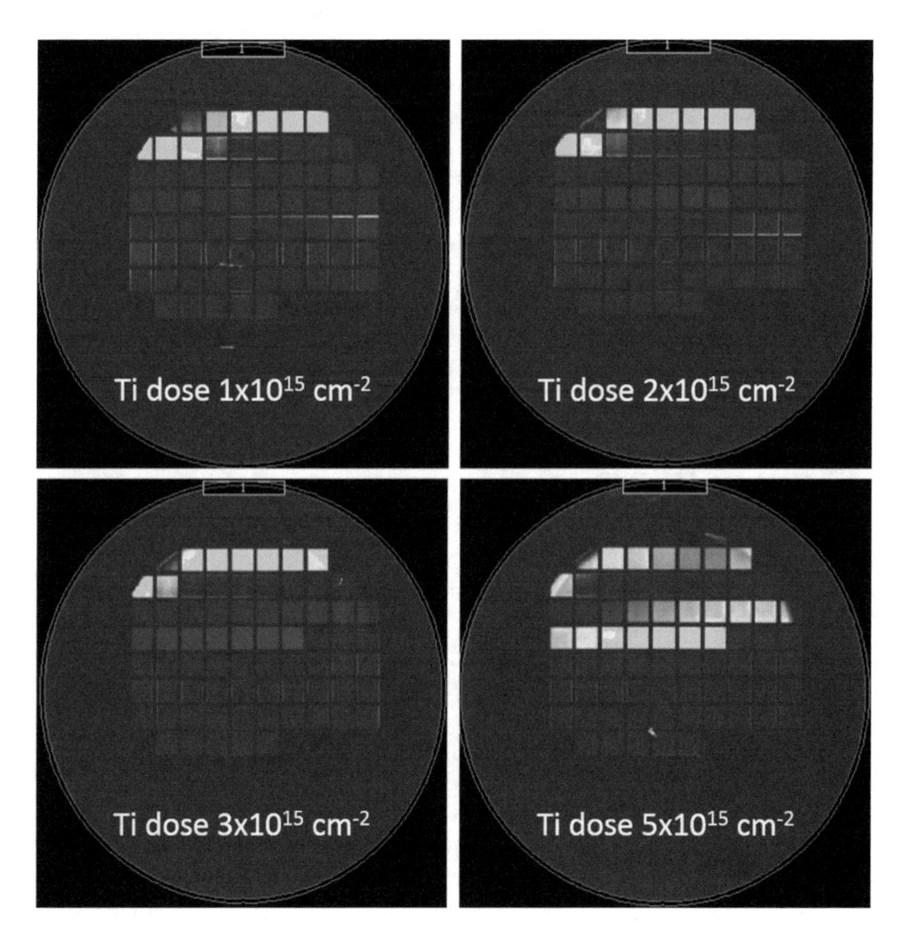

Fig. 5.7 Haze map of four monitor wafers, being each wafer at a different Ti dose. Each square represent a different laser shot of 15×15 mm^2. Haze is increasing from dark grey to white to pink

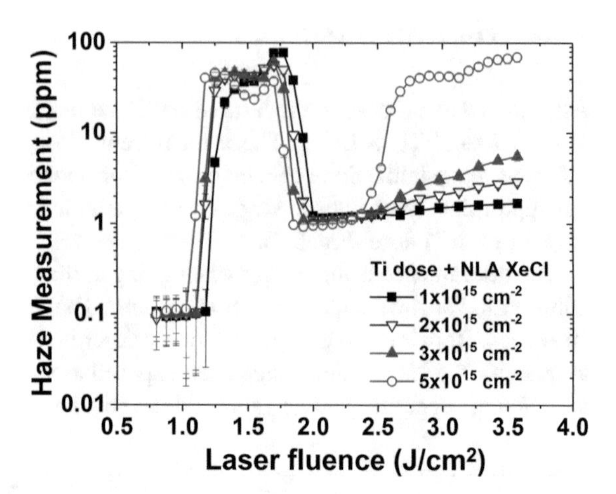

Fig. 5.8 Haze Measurement of Ti implanted samples, as a function of the laser fluence. Results for the four Ti doses are shown in the figure

We observe four regimes in Fig. 5.8. We will take the samples with a Ti dose of 3×10^{15} cm^{-2} as reference to define the different regimes, as we observe how the different regimes start at different laser fluences depending on the Ti dose. The first region goes from 0.80 to 1.10 J/cm^2, characterised by low Haze values. Between 1.10 and around 1.70 J/cm^2, the Haze increases more than two orders of magnitude. The third region is characterised by low values of Haze, between 1.70 and 2.50 J/cm^2. The last region starts at 2.50 J/cm^2 and finishes at the maximum laser fluence achieved by the system, 3.60 J/cm^2, characterised by the increase in the Haze along with the laser fluence, especially visible in the case of the highest Ti dose.

5.5.3 Dynamic SIMS

As in previous chapters, we use SIMS to obtain the Ti concentration profile, aiming to check if the IB formation limit has been achieved. We measured the Ti profile of three out of four Ti doses: 2×10^{15}, 3×10^{15} and 5×10^{15} cm^{-2} on samples with different NLA processes. We decided not to measure samples with the lower dose as the profiles between 2×10^{15} cm^{-2} and 3×10^{15} cm^{-2} were very similar, aiming to save costs and time.

The as-implanted profiles (shown in all graphs by dashed lines) show that there is channelling present in the three Ti doses. The noise floor is located around 10^{16} cm^{-3} in all the samples. As we observed in the SIMS measurements performed in the previous chapter (Fig. 4.2), the melting depth increases along with fluence (estimated as the depth point where the annealed curve goes parallel to the as-implanted one. They should overlap, but since there may be a slight difference inherent of the SIMS measurement process, we estimate the depth using the slope instead). Figure 5.9 upper left shows the Ti profiles of samples laser annealed at 0.95 J/cm^2. The profiles are very close to the as-implanted. Figure 5.9 upper right shows the profiles of samples

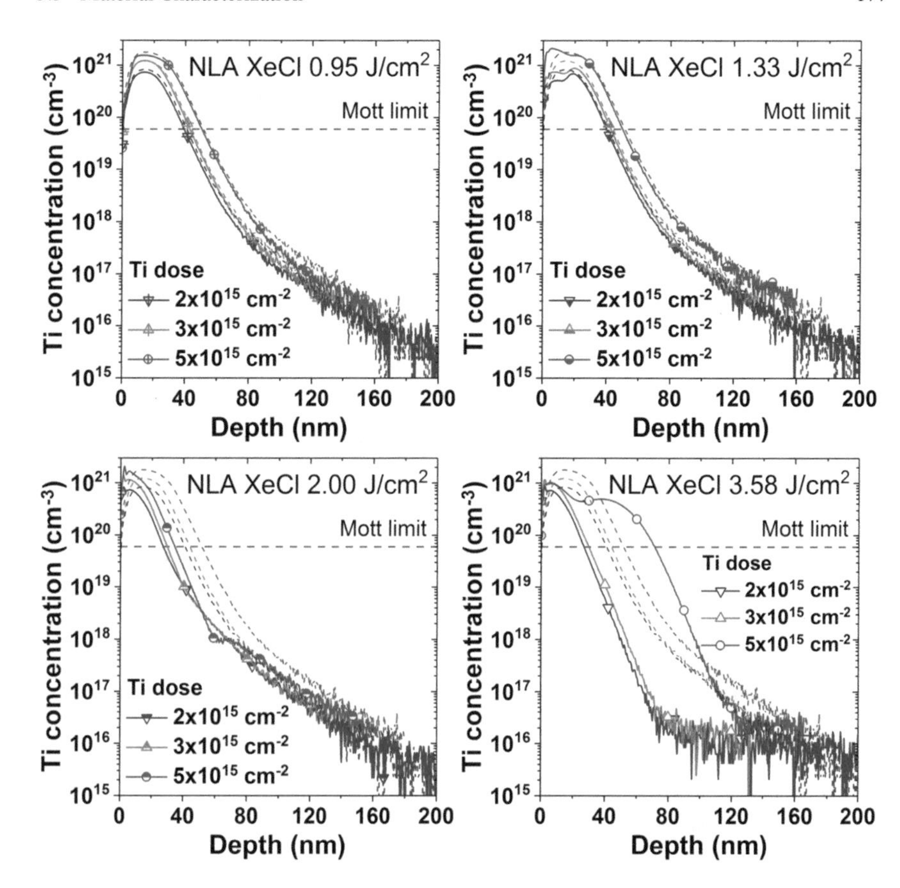

Fig. 5.9 Ti concentration profiles of Ti implanted samples at three different doses. As-implanted profiles are indicated by dashed lines in all the graphs. Each graph shows a different laser fluence used in all the samples: upper left: 0.95 J/cm^2, upper right: 1.33 J/cm^2, lower left: 2.00 J/cm^2 and lower right: 3.58 J/cm^2

annealed at 1.33 J/cm^2, where we observe a redistribution of Ti atoms up to around 25–30 nm for the three Ti doses. Figure 5.9 lower left shows three samples laser annealed at 2.00 J/cm^2, where there is a redistribution of Ti atoms up to between 80 and 120 nm, depending on the Ti dose. Figure 5.9 lower right displays the curves of three samples annealed at the highest laser fluence of this chapter, 3.58 J/cm^2, where all Ti atoms have been redistributed up to 200 nm.

All the samples exhibited Ti concentrations high enough to overcome the IB formation limit, indicated as a horizontal dashed red line. For the first two doses, increasing the laser fluence beyond 2.00 J/cm^2 do not significantly affect the thickness of the Ti supersaturated layer with respect to the thickness at 2.0 J/cm^2. That is not case of the dose of 5×10^{15} cm^{-2}, where we observed that the thickness of the supersaturated layer increased linearly with the laser fluence (featuring a Ti profile with two peaks, as in Fig. 5.9 lower right). The Ti concentration profiles of samples implanted with a dose of 5×10^{15} cm^{-2} resembled the profiles obtained in the previous chapter, with the double ion implantation process and cellular breakdown structures.

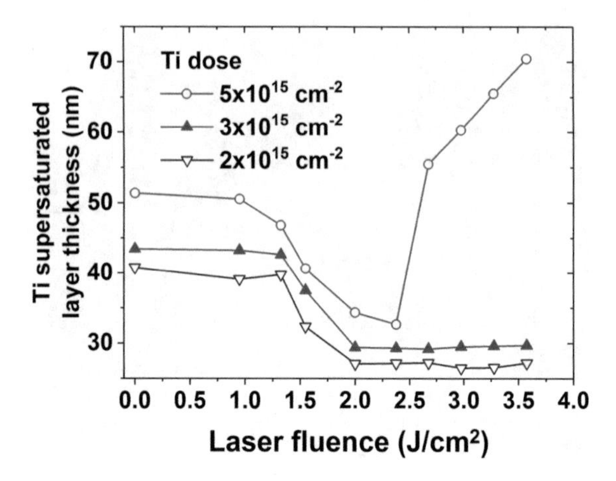

Fig. 5.10 Ti supersaturated layer thickness as a function of the laser fluence for three Ti doses. The thickness is calculated so that the concentration of the layer overcomes the IB formation limit

As expected (Fig. 5.10), the thickness of the Ti Supersaturated Layer (TSL) increases with increasing Ti dose. We observe how the thickness drops down at around 1.55 J/cm^2. The doses of 2×10^{15} and 3×10^{15} cm^{-2} stabilise at lower values while the TSL of the samples with the Ti dose of 5×10^{15} cm^{-2} thicken abruptly for fluences higher than 2.4 J/cm^2.

Another important parameter that can be obtained using SIMS measurements is the retained dose after the laser process. From Chapters 3 and 4 we obtained that there was a loss of Ti atoms after the NLA process. Here, we quantify the retained Ti dose for the samples analysed in this chapter (Fig. 5.11).

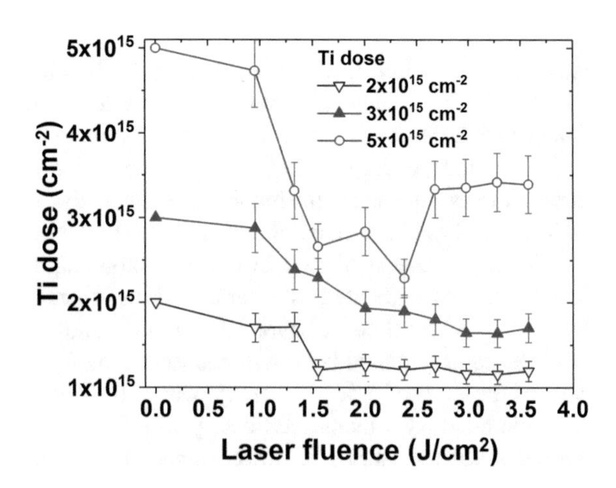

Fig. 5.11 Retained Ti dose as a function of the laser fluence for samples implanted at three different Ti implanted doses

The quantity of retained dose depends on the laser fluence, but generally decreases with increasing laser fluence. The percentage of retained dose is always higher than 54%, which is lower than the values obtained on previous chapters but still high enough to supersaturate Si in the first tens of nanometres. As we observed in the previous figures, the samples having a dose of 5×10^{15} cm^{-2} show a different behaviour. In this case, the retained dose increases from the sample annealed at 2.40 J/cm^2 towards higher laser fluences, with percentages of retained dose around 67%, closer to the values found in the previous chapter using the same XeCl laser. It is yet to determine if the samples implanted at 5×10^{15} cm^{-2} are offering different SIMS results that could indicate the apparition of the cellular breakdown phenomenon, as we observed for samples with the same Ti dose in the previous chapter.

5.5.4 Cross-Section HRTEM

Cross-section HRTEM is used to estimate the crystal quality of the implanted layer before and after the NLA process. We analysed eight samples with different NLA processes at a fixed Ti dose of 3×10^{15} cm^{-2}. Through the analysis of the as-implanted samples we obtain the thickness of the amorphous layer produced by the ion implantation process, around 23 nm. There is also a 2.8 nm thick layer on the surface, which we have identified as the native oxide, by means of Energy Dispersive Spectroscopy (EDS) measurements. We observe three different regimes in the cross-section HRTEM micrographs of these samples.

In the first regime, the amorphous layer coming from the ion implantation process has not been altered, as in the sample annealed at 0.95 J/cm^2, which was very similar to the as implanted one, having the same two layers, 2.8 nm of oxide + 23 nm of a-Si (see Fig. 5.12a, b). The second regime is characterized by recrystallization processes which resulted in non-monocrystalline layers, but nanocrystalline or amorphous instead. In this regime, we found in most samples up to three distinguishable layers: a first amorphous layer 2.8 nm thick, which we identified as the native oxide; a second nanocrystalline layer, containing some randomly oriented nanocrystals; and a third layer made of polycrystalline or poorly crystallized silicon. The sum of the thicknesses of the last two layers was approximately 23 nm, the same value observed on the as-implanted sample. Samples that lie within this regime are shown in Fig. 5.12c, d, where we observed that the thickness of the polycrystalline layer increased along with laser fluence, from 8.4 nm in the case of 1.33 J/cm^2, up to 17 nm at 1.55 J/cm^2. This polycrystalline layer is characterized by a high density of defects and stacking faults. The same way the polycrystalline layer (in contact with the crystalline substrate) thickened along with laser fluence, the nanocrystalline layer atop of it decreased at the same rate. The sample laser annealed with 1.70 J/cm^2 of laser fluence (Fig. 5.12a) is the only one that does not have an amorphous Si layer, with only the native oxide present on the surface, although some stacking faults are still visible. This sample marks the start of the third region, where most of the recrystallized layer is monocrystalline. For the 2.00 J/cm^2 sample (Fig. 5.12b), we found

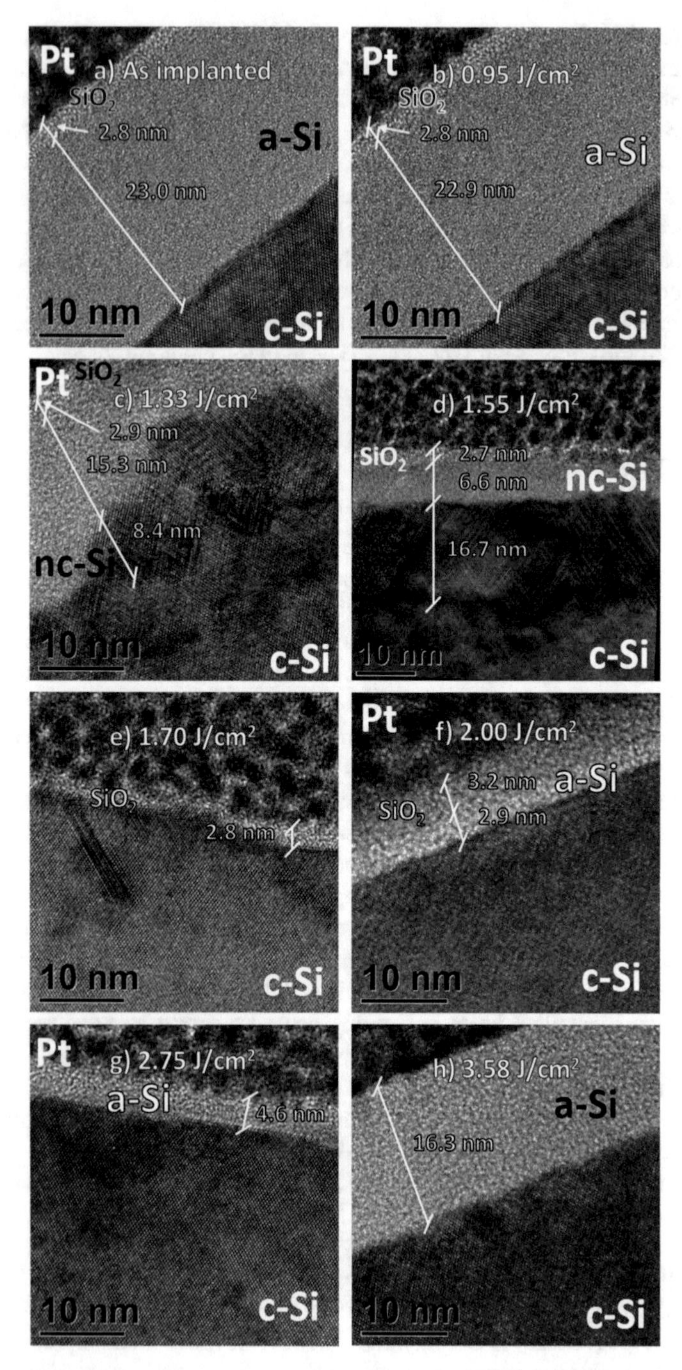

Fig. 5.12 Cross-section HRTEM micrographs of samples Ti implanted with a dose of 3×10^{15} cm^{-2}, with different laser processes. **a** As-implanted, **b** 0.95 J/cm^2, **c** 1.33 J/cm^2, **d** 1.55 J/cm^2, **e** 1.70 J/cm^2, **f** 2.00 J/cm^2, **g** 2.75 J/cm^2 and **h** 3.58 J/cm^2

an amorphous layer around 7 nm thick. EDX measurements showed that the first 2.8 nm are due to the presence of oxygen, while the rest of the layer is composed of titanium and silicon, being around 3.5–4.5 nm thick. Similar a-Si layers are observed for 2.75 and 3.58 J/cm^2 with 5.3 nm and 17 nm of thickness respectively (Fig. 5.12c, d).

5.5.5 *Atomic Force Microscopy*

A few lines back, we described the use of the Haze to obtain a first estimation of the surface roughness, which could be potentially used to identify the different crystallisation regimes. AFM is shown here to validate the results obtained by HM. As in the previous sub-section, we only analyse the samples implanted at a dose of 3 $\times$ 10^{15} cm^{-2}. We used the AFM instrument described in Sect. 2.3.3 in tapping mode, took 5 $\times$ 5 μm scans and measured the RMS value of the roughness (Fig. 5.13).

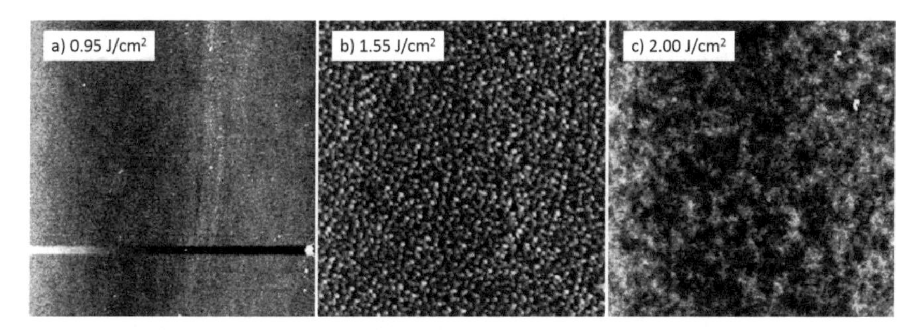

Fig. 5.13 AFM scans of 5 $\times$ 5 μm of samples Ti implanted at a dose of 3 $\times$ 10^{15} cm^{-2}, laser annealed at 0.95 J/cm^2 (left), 1.55 J/cm^2 (centre) and 2.00 J/cm^2 (right)

The sample annealed at 0.95 J/cm^2 of laser fluence shows a rather flat profile. Increasing the laser fluence up to 1.55 J/cm^2 modifies the surface, showing a granular-like structure. The granular structure seems to disappear if the laser fluence is further increased to 2.00 J/cm^2. In order to quantify the surface morphology, we represent the RMS roughness of the AFM scans. Along with the RMS roughness we show also the Haze curve of samples with the same Ti dose to further examine their possible relationship (Fig. 5.14).

We found four distinguishable regimes in AFM measurements. AFM RMS roughness value is low, starting from the as-implanted sample up to 1.10 J/cm^2 (the limit of the first regime). The values are slightly above the minimum resolution of the AFM microscope, leading to high uncertainties. After that, roughness increases abruptly up to around 2 nm (second regime). Beyond 1.70 J/cm^2, we defined the third regime in which the roughness drops down. There is a region with minimum values around 2.30 J/cm^2, after which roughness increases again but at a lower rate, delimiting

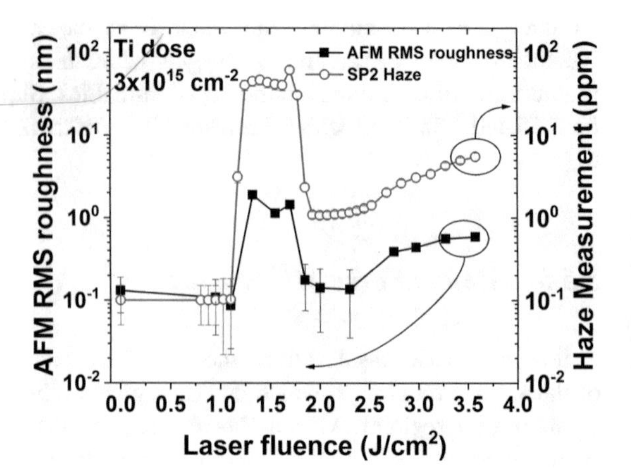

Fig. 5.14 AFM RMS roughness (left Y-axis) and Haze Measurements (right Y-axis) as a function of the laser fluence for samples Ti implanted at a dose of 3×10^{15} cm^{-2}

the fourth regime. We observe the same intervals and trends in Haze measurements, which evidences the qualitative relationship between the two techniques, AFM and HM, as previously referenced by other authors [12]. Judging from the available experimental data and their related uncertainties, Haze measurements may be more sensitive to slight variations, as compared to AFM RMS value.

5.5.6 *Sheet Resistance*

Sheet resistance measurements provide information about the conduction processes inside the TIL. As we described in the previous chapter, they may offer indirect information about the crystal quality; monocrystalline structures are usually better conductors than polycrystalline structures. The measurements shown here were performed at wafer level using the aligned four-probe equipment described in Sect. 2.4.4. We took five measurements inside each laser annealed die, avoiding the borders of each die. The average value is represented.

There are clearly two different regimes in the sheet resistance measurements: a regime of low laser fluences, where the sheet resistance is high, in the order of 10^5 Ω and Ti dose does not seem to have an influence, and the regime of high laser fluences, where the sheet resistance drops down between 10^3–10^4 Ω, and there is a clear dependence with the Ti dose. The sheet resistance decreases in this region with increasing Ti dose, in line to what we measured in the previous chapter with double implanted samples using the van der Pauw configuration. At laser fluences higher than 2.40 J/cm^2, the samples with the highest Ti dose shows an increase in the sheet resistance, not observed in any of the other doses. Note that the resistivity of the substrate is 15 Ω·cm, which should lead to sheet resistance values of around 200 Ω taking into account that the wafer thickness is 750 μm. The measured value in the outside of the implanted region was 360 ± 20 Ω. The value is considerably lower than

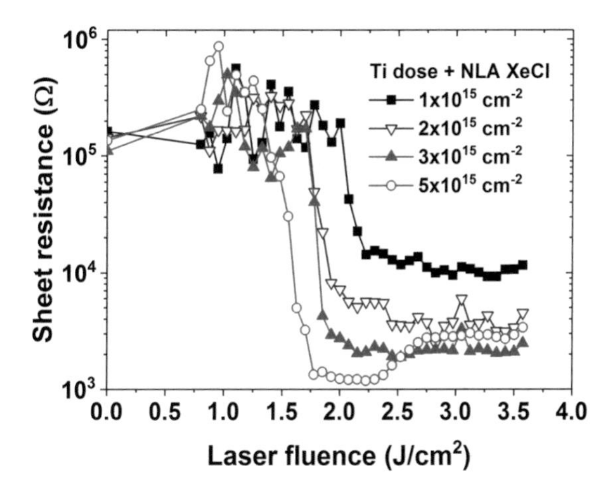

Fig. 5.15 Sheet resistance as a function of the laser fluence for samples implanted at four Ti doses. The resistivity of the substrate is 15 Ω·cm

the ones represented in Fig. 5.15 for Ti implanted samples and close to the expected value of 200 Ω using geometrical calculations, considering that the thickness of the wafer is close to the separation between probing points and correction factors should apply.

The transition between the high and low sheet resistance changes along with the Ti dose, as we observed with Haze measurements.

5.5.7 *Discussion on Material Properties*

According to the different experimental results shown in the previous sub-sections, the experimental evidences point out to an explosive recrystallization process, similarly to what we observed in the previous chapter, when laser annealing Ti implanted samples with the same laser equipment. In this chapter, we used different techniques than in the previous chapter, which corroborated the recrystallization regimes briefly discussed in Chapter 4. With the new techniques used in the internship we could better delimit the different recrystallization regimes.

TRR measurements may provide key information to identify the recrystallization processes as a function of the laser fluence. The increase of the reflectivity has been linked to an increase of the temperature of Si samples at the surface [13], until a saturation value is reached, which is consistent with the relatively constant temperature of a molten layer. When the reflectance peak reaches saturation, it is no longer a proper peak, but a "plateau". In fact, the width of the plateau is used to probe the time in which the layer stays molten [11]. When the laser energy density is sufficiently low, the top surface does not melt. Thus, the TRR curve show little variations. Once the temperature increases up to a certain threshold, we observe a first peak. If we increase further the laser fluence, we observe two different peaks in Ti implanted samples, labelled as α and β_1 in Fig. 5.5. Thus, a two melting process for

Ti implanted Si samples could be inferred, one involving the α peak and the second involving the β_1 peak/plateau. On the other side, we could infer only one process for bare Si substrates, associated with the β_2 peak/plateau.

As we have seen on HRTEM micrographs, the Ti ion implantation process produces an amorphous layer on the surface, around 23 nm thick. As described in Sect. 4.3.1, the melting and solidification processes of an amorphous layer are considered to be very fast, as the molten Si coming from an amorphous phase is an undercooled liquid [14]. The short duration of the α peak, which does not depend on the laser fluence, points out to its possible origin in the melting and solidification process of the amorphous silicon. Besides, its lower reflectivity value, with respect to the β_1 peak, is consistent with the lower melting temperature of amorphous silicon [15]. Other evidences support our proposition linking the nature of the α peak and the melting process of the amorphous layer: the non-implanted wafer did not exhibit the α peak, and samples annealed with two laser shots in the same area only showed the α peak in one of the TRR curves. In this last group of samples, we examined the presence or absence of the α and β_1 peaks, at fluences close to the frontier between the "Sub-melt" and "First melt and solidification" regimes.

In effect, in Fig. 5.16 left, in the first NLA process we used a laser fluence in the sub-melt regime, at 1.03 J/cm^2, which lead to a rather flat TRR profile, with no observable abrupt peaks, indicating that the laser fluence was not high enough to possibly melt the amorphous layer in the surface. During the second laser process at 2.40 J/cm^2, framed in the "high energy" regime, the curve displayed the α peak and β_1 plateau, indicating that both melting processes took place. In Fig. 5.16 right, the laser fluence of the first shot, at 1.10 J/cm^2, belongs to the "First melt and

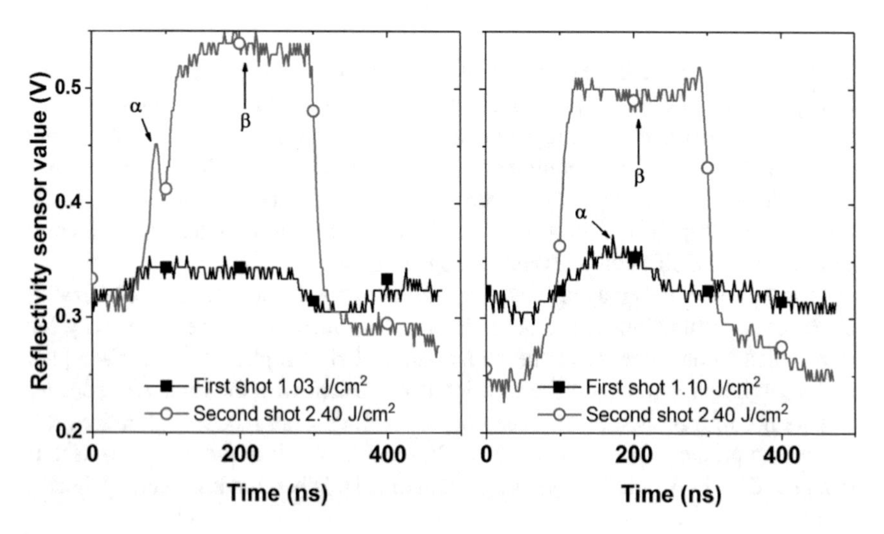

Fig. 5.16 Reflectivity sensor value as a function of elapsed time during NLA of Ti implanted samples. In the left figure, the first laser shot was fired at a fluence of 1.03 J/cm^2. Then, a second laser shot of 2.40 J/cm^{2}annealed the same surface. In the right figure, the first laser shot was fired at a fluence of 1.10 J/cm^2, and the second one also at 2.40 J/cm^2

solidification" regime. The curve shows the α peak at around 170 ns from trigger, which may indicate that at least part of the amorphous layer was melted. During the second NLA process, at a fluence of 2.40 J/cm^2, we could only observe the β_1 plateau in the TRR curve. This could indicate that the amorphous layer was melted during the first NLA process, possibly reordering the atomic structure of the annealed layer into nanocrystalline or polycrystalline silicon. Thus, the second laser process only melted Si in monocrystalline or polycrystalline phases, associated to the presence of the β_1 plateau. Thus, the apparition of the α peak only once in several laser shots (Fig. 5.16) is compatible with the expected behaviour under the explosive recrystallization assumption.

Other evidences that further support the origin of the α peak are the cross-section HRTEM micrograph of the sample laser annealed at 0.95 J/cm^2. This sample showed the same amorphous structure, indicating that no melting took place during the laser process. Its Ti concentration profile (by SIMS) did not show any change with respect to the as-implanted samples. Besides, its TRR curve did not show the α peak. If we examine the samples laser annealed at 1.33 and 1.55 J/cm^2, we observe the presence of the α peak in the TRR curve, while at the same time we observe in their micrographs that recrystallization within the amorphous layer took place. The SIMS profile is consistent, where there was a redistribution of Ti atoms up to a depth of around 25 nm (see Fig. 5.9b), close to the thickness of the amorphised layer by the ion implantation process.

It seems clear that the origin of the β_1 and β_2 peaks is related to the melting process of crystalline silicon. Once the layer is melted, there is no significant increase in the temperature, which is indirectly measured by a constant saturation reflectance in β_1 and β_2 peaks/plateaus. In non-implanted monocrystalline Si, increasing the laser fluence beyond the threshold increases the duration of the melt, in a linear relationship, as described by other authors [6]. Our experimental results of the β_2 peak/plateaus that appeared in non-implanted crystalline substrates (Fig. 5.6), agree with the linear trend described in the previous reference. The fact that the β_1 peaks follow the same linear trend supports our hypothesis linking the origin of the β_1 and β_2 peaks to the melting of the monocrystalline part of the substrate. More evidences further agree with our assumptions. At 1.70 J/cm^2, the TRR curve shows a reflectance value at the β_1 peak close to saturation value, indicating that part of the crystalline layer may have been melted. Its HRTEM micrograph shows a monocrystalline structure, although some defects are still visible. The Haze and AFM RMS roughness values show a drastic reduction of the roughness, which would be consistent with a smooth crystalline layer. Besides, the sheet resistance values show a drastic reduction at the same laser fluence; conduction through a monocrystal is easier than through a polycrystalline structure.

Although linked to the same physical phenomenon, there are small differences between the β_1 peak/plateau and the bare Si β_2 peak/plateau. First, the minimum laser fluence required to see the peaks: Ti implanted samples showed their β_1 peak at laser fluences lower than the non-implanted reference, and, according to Fig. 5.6, this minimum threshold also depends on the Ti dose: higher Ti doses show the β_1 at lower laser fluences. This could be explained by the amount of damage that is still present in the crystalline phase under the amorphous implanted layer. Although the phase is crystalline, some defects are expected closer to the amorphous-crystalline interface (the EOR defects [16]). The induced damage by the ion implantation could lower the threshold value necessary to melt the layer.

The reflectivity saturation value of the β_1 plateaus is much lower than for β_2 plateaus, which may indicate a different light absorption process on Ti implanted samples, as compared to bulk Si samples, at $\lambda = 635$ nm, during the melted phase. In the previous chapter, we estimated the complex refractive index of the TIL, using the four-layer model (Fig. 4.16) at room temperature. The differences observed in the reflectivity of the molten phase are consistent with a change in the complex refractive index due to the high Ti concentration, as we observed in the previous chapter on samples at room temperature. Since n and k are interdependent it is hard to extract more conclusions in the molten phase.

Another difference between β_1 and β_2 peaks is the FWHM: Ti implanted samples remained in the liquid state longer time than non-implanted Si, which happens at all the laser fluences used in this work. Besides, samples with higher Ti dose stayed melted longer. Longer melting times could be related to lower thermal conductivities found on heavily implanted Si layers, mainly due to phonon scattering processes. Other authors pointed out that the thermal conductivity is severely influenced by the presence of impurities, vacancies or stacking faults, among others [17]. This affirmation would be consistent to what we have experimentally observed.

Based on the available data, we have labelled four different regimes of crystallization for Ti supersaturated Si layers, taking as reference the samples with a Ti dose of 3×10^{15} cm^{-2}. As commented before, Ti dose influences the threshold values between different regimes.

Sub-melt: up to 1.10 J/cm^2. The amount of energy delivered is too low to start melting the amorphous phase. No observable features are seen on TRR curves. HRTEM micrographs are similar to the as-implanted sample. No redistribution of Ti atoms is observed by SIMS. The roughness, measured by HM and AFM is similar to the as-implanted samples. Sheet resistivity is high.

First melt and solidification: from 1.10 J/cm^2 up to 1.40 J/cm^2. One peak (α) is observed in TRR measurements. Only the amorphous layer is expected to be partially or totally melted, then subsequently solidified into polycrystalline or amorphous material. It is the so-called explosive recrystallization phenomenon. HRTEM shows nanocrystals in the implanted layer. SIMS profiles show a redistribution of Ti atoms up to the a-c interface. Haze and AFM show an increase in the roughness compatible with the presence of nanocrystals growing from the a-c interface. Sheet resistance is still high.

Second melt and crystallization onset: between 1.40 and 1.70 J/cm^2. Laser energy density starts to be high enough to lead to a first melt of the amorphous layer as in the previous regime. The layer solidifies quickly, following the explosive recrystallization phenomena, which starts to melt immediately after as the laser is still annealing the sample. This second melt is responsible for the appearance of the β_1 peak. HRTEM shows an increase in the size of the nanocrystals and polycrystals growing from the a-c interface. Haze and AFM show high roughness values and sheet resistance is still high.

High energy: for laser fluences higher than 1.70 J/cm^2. The second melt is propagating deeper and deeper in the structure, exceeding the initial a-c interface, as the laser fluence is increased. This results in the saturation of the reflectivity signal at its maximum value, around 0.52 V for Ti implanted samples, and a β_1 peak that evolves into a plateau with increasing width. HRTEM show a monocrystalline layer in most of the implanted volume. SIMS profiles show a redistribution of Ti atoms beyond the a-c interface. Haze and AFM show a decrease in the roughness, reaching values close to the as-implanted although slightly higher. Sheet resistance is low, accounting from a better conduction through a monocrystalline substrate.

We observed that for Ti implanted samples, Haze values were always higher than the as-implanted ones. It is well known that after a NLA process some roughness may be induced on the surface [18]. The size of these induced structures is very small, but important enough to be detected by either Haze or AFM measurements. The induced peaks have an average height of 2–3 nm while having a width of a few tens of nanometres for almost every laser energy density used, once the amorphous layer has been molten, according to AFM. This effect is particularly visible for samples showing polycrystalline behaviour. Taking a look at TEM micrographs (Fig. 5.12) the small crystals seem to push upwards the remaining nanocrystalline layer, producing visible protuberances that could be the origin of the grain-like surface seen on AFM roughness maps (Fig. 5.13). In conclusion, Haze for Ti implanted samples comes from both polycrystalline behaviour and induced roughness, with regimes in which no polycrystals are present but higher Haze values than the as-implanted sample are measured. Nevertheless, when Haze is maximum, polycrystals are found, so this information could be used to find out the different regimes in which the material is well crystallized.

Relative to crystal quality of supersaturated layers, several works [19, 20] have stated that it is very difficult to achieve Ti supersaturation on Si without the appearance of CBD structures phenomenon. Here, we could not observe this phenomenon on any of the examined samples having a Ti dose of 3×10^{15} cm^{-2}, which is positive as we expect to reduce the possible non-uniformities of the material, a desired characteristic in materials with optoelectronic applications. It is yet to determine if the samples with a Ti dose of 5×10^{15} cm^{-2} show CBD structures, as we observed using the same XeCl laser but on double Ti implanted samples described in the previous chapter.

In the high energy regime, we measured several unexpected behaviours. At laser fluences higher than 2.40 J/cm^2, we observed an increase in the Haze that we linked to the unexpected presence of a regrown amorphous layer on the surface, according to TEM. The micrographs showed that the thickness of the amorphous layer increased along with laser fluence (Fig. 5.12f–h). The recrystallization theory predicted an epitaxial growth from the crystalline substrate once the threshold for crystalline silicon melting was achieved, with little to no difference in the crystal quality of samples annealed with fluences over the threshold. This is what we have been observing for the past years in our research group with Ti supersaturated Si, as described in Chapters 3 and 4. Other works have documented the regrown of amorphous layers on supersaturated Si but with different laser processes, namely femtosecond laser annealing, and with other chemical species such as S, Se or Te [21, 22].

We have considered multiple options in order to explain the origin of the regrown amorphous layer. We proposed a first option, which would involve oxygen in the lattice coming from the atmosphere or the native oxide. However, we could not trace oxygen along the whole regrown amorphous layer using EDX. We considered a second hypothesis in which an excessive amount of Ti atoms close to the surface, either as an agglomerate of Ti clusters, or as part of a silicidation process could explain the presence of the amorphous layer, but we could not detect Ti using EDX on STEM on it. Taking into account that EDX has a minimum sensitivity value of around 1–2% of atomic percentage, not being able to detect Ti on the regrown amorphous layer at high laser energy densities led us to discard that the amorphous layer was composed of Ti agglomerates, which would require considerably higher concentrations.

Lastly, we propose that the solidification speed could have been too high during certain intervals of the recrystallization process, higher than 15 m/s, as discussed in Sect. 4.3.1. Under such circumstances, the solidified layer would recrystallize as amorphous. However, we used the same laser process to anneal Ti implanted samples in the previous chapter, and none of the samples analysed there showed amorphous layers for laser fluences in the high energy regime. The main difference was the energy of implantation, which drastically affects the thickness of the amorphised layer: 23 nm in this set of samples compared to around 155 nm in samples with double Ti implantation, depending on the dose. During the first moments of light absorption, most energy is trapped in the amorphised layer. That would imply that higher temperatures would be expected in the thinner layer. Since speed of recrystallization depends on the gradient of the temperature, it is expected that, since the temperature is higher, the solidification speed would also be higher [23]. This possibility is consistent with what we observed from available data in the high energy regime, as the thickness of the regrown amorphous layer is increasing along with the laser fluence, which would relate to higher speeds of solidification. Also, it has been found in the literature that melting thin amorphous films using excimer lasers could be more challenging as compared to thicker films [24] due to irregularities in the light absorption and subsequent melting processes.

Regarding the sheet resistance measurements, they show similar results to those shown in Sect. 4.3. The sheet resistance of Ti implanted samples is at least two orders of magnitude higher than the value expected for the substrate alone, set to around 15 Ω·cm, which points out to a decoupling process of the substrate from the TIL at room temperature. According to the findings of Garcia-Hemme et al. [25] the photoresponse of the TIL increases when the samples is at temperatures close to what we have labelled as the second transition. Also, in the regime of high energy, the sheet resistance decreased with increasing Ti dose, which is consistent with what we measured in Sect. 4.3 at room temperature.

Another important aspect related to the fabrication of the Ti supersaturated material is that we did not found any single crater on any sample analysed in the 300 mm wafers used in this chapter, even though more than 2500 cm^2 were laser annealed on Ti implanted areas. This further evidences the origin of the craters in the deposition of solid particles before the NLA process. 300 mm wafers in an industrial route are always kept in the cleanest environments. Their level of cleanliness is monitored after each fabrication step, to assure it lies within specifications. Further proof are the Haze maps shown on Fig. 5.7. Craters featured sizes in the order of microns, and should be clearly visible in the Haze maps.

Wrapping up the results from the material characterisation, we were able to check that the condition of supersaturation was achieved. We also obtained the recrystallization regimes at the four Ti doses used in this chapter, regimes which are consistent with the explosive recrystallization model. After the analysis, we identified the best laser conditions for device fabrication, which should lie between 1.70 and 2.40 J/cm^2 approximately.

5.5.8 *Material Characterisation on Patterned Wafers*

After the analysis of the monitor wafers, we show part of the characterisation process on patterned wafers. One of the main concerns at the moment of the NLA process was the possible effect on the pixel structure. The high temperature gradients between the molten and the solid phase could affect the pixel integrity, in particular the integrity of the trenches filled with silicon dioxide, which act as isolation barriers (Capacitive Deep Trench Isolations, CDTI structures). We used an optical microscope to observe the pixel structure right before and after the NLA process on devices with a Ti dose of 5×10^{15} cm^{-2}. The results are shown below:

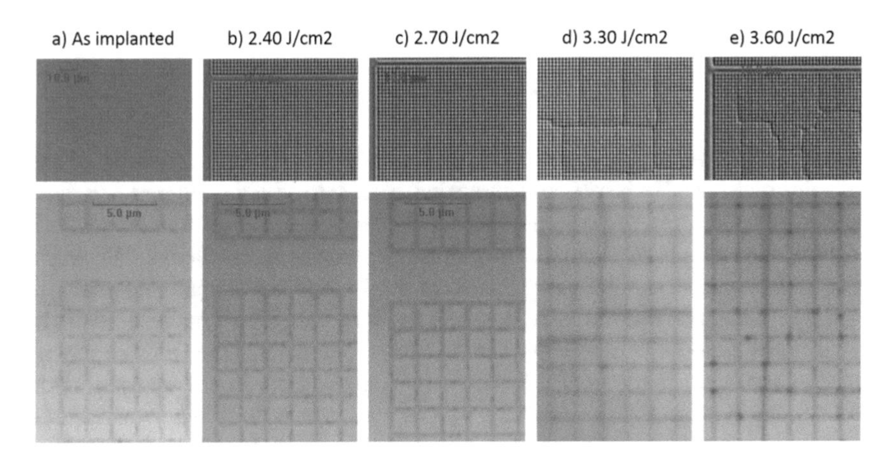

Fig. 5.17 Photography through an optical microscope of the backside of the pixel matrix. The upper row of pictures is taken at 20X magnification. The lower row was taken at 150X magnification

We see that annealed devices showed more contrast in the pictures, making the pixel matrix more visible as compared to the as-implanted. However, for fluences higher than 2.70 J/cm^2 there seems to be an incompatibility problem between the NLA process and the pixel structure. Taking a look at pictures d) and e) in Fig. 5.17 we see marked paths, following the CDTI structure between pixels. In figure e) particularly, it may seem that the dark dots seen in the as-implanted sample are darker than the rest of the annealed samples. The highlighted paths seem to break some pixels, as seen in bottom pictures in Fig. 5.17e). The presence of these defects must be taken into account when examining the properties of the fabricated devices. It seems that devices with fluence equal or lower than 2.70 J/cm^2 won't show this problem.

5.6 Device Characterisation

The characterisation of the Ti supersaturated material, described in the previous section, helped to determine the range of laser fluences in which we could expect good crystal quality on the TIL. In this section, we finally measure the fabricated devices and analyse the results.

There are two possibilities when it comes to characterise the fabricated devices: processing at wafer level (Wafer Level Characterisation, WLC) and die-per-die characterisation, where dies were cut from the wafer, wire-bonded to a chip carrier and measured individually. Only samples with a Ti dose of 2×10^{15} cm^{-2} were diced and mounted on chip carriers for individual testing. The rest of the doses were analysed using WLC.

5.6.1 Testing the Read-Out Integrated Circuitry

In Sect. 5.5.8, we evaluated the effect of the laser process in the integrity of the backside of the pixel (where Ti was implanted, the frontside of the wafer as described in Fig. 5.2). However, the laser process could affect the functionality of the ROIC, as high temperatures are expected in volumes close to the melted layers. The pixels used in this chapter are 3.5 μm thick, so an increase of the temperature could be expected at the frontside of the pixel, where the ROIC and transistors are located. In order to see if the MOS transistors were affected, we measured the Drain current as a function of the voltage applied between the Drain and the Source by using some test structures present in the die using a probe station at wafer level. We measured the worst-case scenario, which would correspond to the highest Ti dose, 5×10^{15} cm^{-2}, and the highest laser fluence, 3.60 J/cm^2 (Fig. 5.18).

Along with the Ti-implanted die, we measured also a non-implanted die submitted to the same laser process and a non-implanted, non-annealed Si die. The results are identical in the linear and saturation regions of the transistor. Only a slight difference

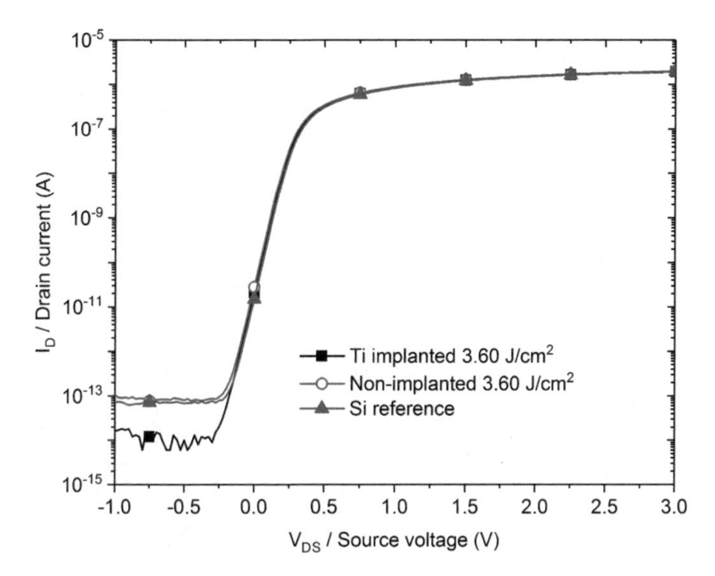

Fig. 5.18 I_D-V_{DS} curve for the READ transistor of a pixel matrix Ti implanted at a dose of 5×10^{15} cm^{-2}, laser annealed at 3.60 J/cm^2, a reference pixel matrix annealed at the same laser fluence and non-implanted, non-annealed die

is seen for negative voltages in the off-state current, where the Ti implanted sample exhibited lower values. We extracted the threshold voltage of the transistors of dies having different Ti doses and NLA processes to further analyse if the NLA process could have affected the doping of the transistors. The results lie within specifications of the reference sample, set to 0.300 ± 0.025 V in all cases. Therefore, the transistors were not affected by the NLA process.

5.6.2 Camera Acquisition Pictures

Individual dies mounted on chip-carriers are connected to a motherboard, which both polarises the device and read the data coming from it. During the first tests, the pixel matrix is connected to the computer while being illuminated. The computer processes the data coming from each pixel and forms an image. This first image can be used to examine the quality of the pixels and the uniformity at first sight, without entering into data processing or statistics.

The camera acquisition pictures show the distribution of the pixels between different families. For simplicity, only six families of pixels were shown, although there were 10 versions. The dark stripe on the left of each picture is composed of pixels deliberately covered. They are used to measure the dark current and provide a reference frame for the illuminated pixels. The upper row of Fig. 5.19 shows samples non Ti implanted, with increasing laser fluence from left to right. We observe how the

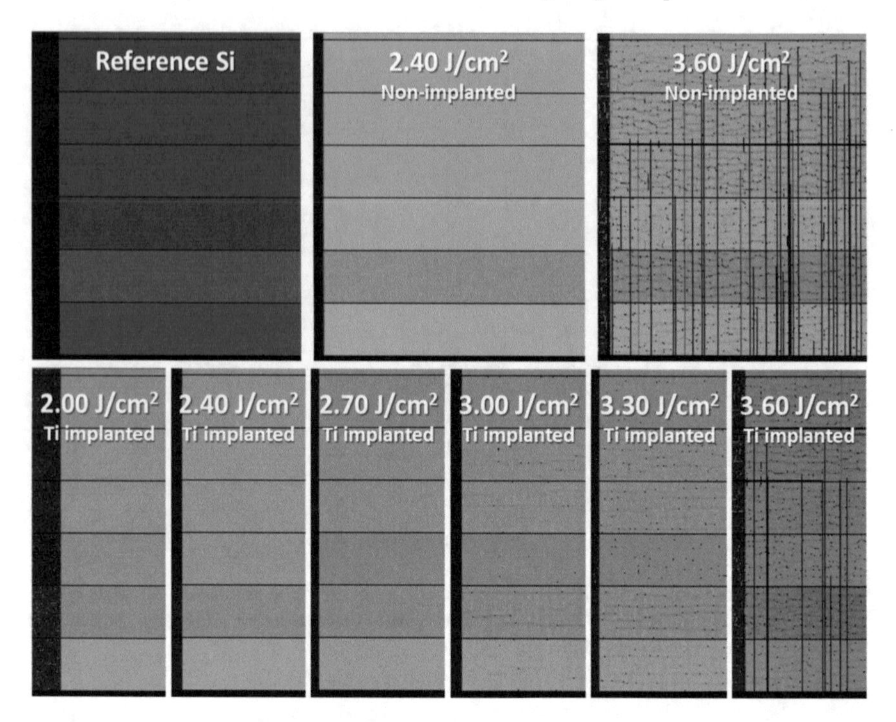

Fig. 5.19 Camera acquisition images obtained by the pixel array in illumination mode. First row contains non-implanted dies with different laser fluences, while the second displays the Ti implanted dies at a dose of 2×10^{15} cm^{-2}. Integration times between samples may vary, so the brightness shall not be used to compare between samples

highest laser fluence affects both the covered and illuminated pixels, obtaining dark grey patterns that remind to those obtained from the optical microscope (Sect. 5.5.8). A similar effect is seen on samples that were Ti implanted. In the case of Ti implanted samples, it is remarkable how the covered pixels showed higher dark current values (observable as "hot" or "white" pixels due to their clearer appearance) in the case of 2.00 J/cm^2. The amount of white pixels decreased abruptly at higher laser fluences, up to 3.00 J/cm^2. The pernicious effect of high laser fluences is visible beyond 3.00 J/cm^2 on Ti implanted samples as well. Comparing Ti implanted and non Ti implanted dies we conclude that the deterioration of the pixel matrix is caused solely by the NLA process when laser fluences higher than 3.00 J/cm^2 are used.

Up to this point, we can assure, based on the experimental results, that all the extra fabrication steps, including the Ti ion implantation and the NLA process, are fully compatible with the fabrication route of a pre-commercial CMOS Image Sensors, provided that the laser fluence is lower than 3.00 J/cm^2. This result is a crucial first step towards the integration of the Ti supersaturated Si layers, opening the possibility to explore other pixel structures in collaboration with STMicroelectronics. The contamination levels of Ti, as measured by Total Reflection X-ray Fluorescence (TXRF), along the whole fabrication process, was proven to lie within specification inside STMicroelectronics and CEA-LETI fabs, which guarantees the compatibility of Ti implanted wafers with the rest of processes inside their facilities.

5.6.3 *Wafer Level Characterisation*

Most of the wafers were measured using the WLC process, where devices are measured using probe stations in a fully automatized instrument. Results from the non-implanted and Ti implanted dies are obtained as a function of the laser fluence and the Ti dose. The number of parameters measured in the WLC is limited as compared to individual die characterisation. The values presented here refer to the same pixel version, and the average value of the whole wafer per each Ti dose is shown. The uncertainty is the standard deviation (of > 10^7 pixels). Wafers with Ti doses of 10^{15}, 3×10^{15} and 5×10^{15} cm^{-2} were measured. All measurements were taken at 25 °C (Fig. 5.20).

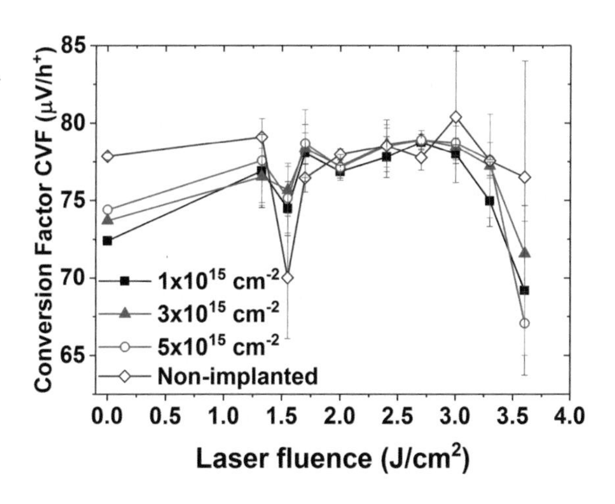

Fig. 5.20 Conversion Factor as a function of the laser fluence for Ti implanted and non-implanted pixel matrices

The first measurement done at WLC is the conversion factor (CVF), which is necessary to calculate the rest of parameters, including the EQE. Except at low fluences, in the first crystallisation and solidification regime, the CVF of the annealed dies in both Ti implanted and non-implanted samples show the same values, within the uncertainty of the measurement, value that is close to the non-implanted, non-annealed die, set to 77.7 $\mu V/h^+$. This result is positive, as it means that the main mechanisms inside the pixel that converts the stored charges (photogenerated holes in this case) to a measurable voltage were not affected by the Ti implantation nor the NLA process. This is not the case for laser fluences higher than 3.00 J/cm^2, where CVF decays for Ti implanted samples. This further confirms that the Ti supersaturated

Si layer has been successfully integrated in the pixel fabrication route. Measuring incorrect values of the CVF would rend the devices useless.

The next key parameter measured at WLC is the dark current (Fig. 5.21).

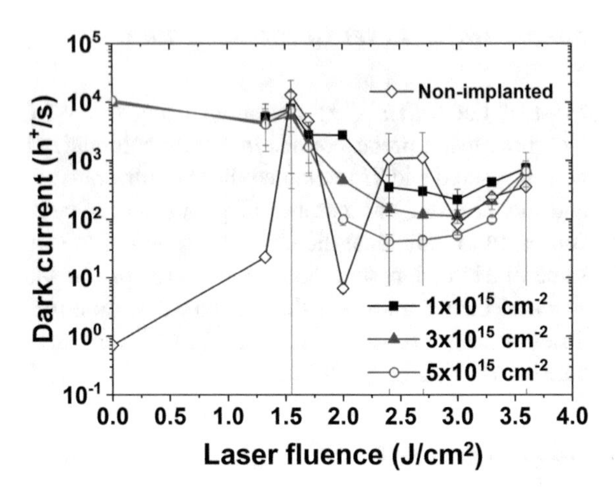

Fig. 5.21 Dark current as a function of the laser fluence for Ti implanted dies at different doses and non-implanted dies

The dark current is a key parameter that affects the performance of the pixels, especially at low light conditions. The standard reference pixel (non-implanted, non-laser-annealed) exhibits a dark current, at 25 °C, of less than one hole per second (less than 0.2 aA). Following the non-implanted curve, we observe how the NLA process drastically increases the dark current to levels in the order of 10^3 h^+/s. The Ti implanted samples exhibit currents in the order of 10^4 h^+/s in the as-implanted case. Once the high energy regime has been achieved (where good monocrystal quality is expected), the dark current decreases to levels between 80–800 h^+/s, depending on the Ti dose. In this regime, we see how increasing the laser fluence beyond 3.00 J/cm^2 increases the dark current, and that increasing the Ti dose decreases the dark current. Vertically pinned-photodiodes, another standard pixel structure, usually offers noise levels in the order of 40-120 electrons or holes per second [When there are eno], which are not far from the values reported here for Ti implanted dies.

According to these results, the NLA process would be responsible of the increase in the dark current in more than two orders of magnitude. The Ti dose influences the dark current, as well as the specific value of the laser fluence. The best devices in terms of dark current would be those with Ti doses of 5×10^{15} cm^{-2}, laser annealed at a fluence of 2.40–2.70 J/cm^2. We do not know at the moment the origin of the increased noise after the laser process. We expect to perform TEM on non-implanted and laser annealed dies to further understand the results.

Finally, we show the EQE at several wavelengths, only for samples having a Ti dose of 3×10^{15} cm^{-2}. Unfortunately, the optical systems available at ST at the WLC were limited to wavelengths between 400 and 1000 nm, which corresponds to energies higher than the bandgap (Fig. 5.22).

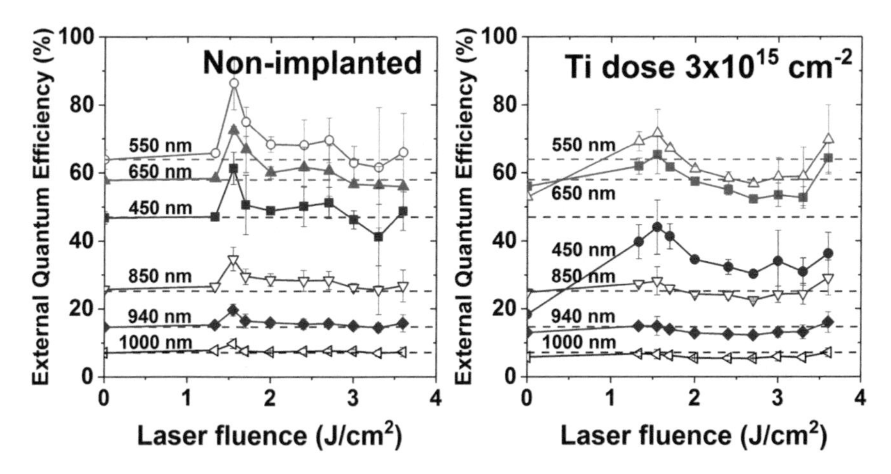

Fig. 5.22 External quantum efficiency as a function of the laser fluence at different wavelength (coded in colours). Left: non Ti implanted samples. Right: samples Ti implanted at a dose of 3×10^{15} cm^{-2}. Horizontal dashed lines indicate the EQE of the reference die (non-implanted, non-annealed)

First, we analyse the results on non-implanted dies. The EQE increases if the laser fluence is located between 1.55 and 3.00 J/cm^2 for all the wavelengths, which is an unexpected result. However, the values close to 1.55 J/cm^2 should be taken cautiously, as those laser fluences led to the highest dark values, which could distort the EQE measurement: dark current is mathematically subtracted from the reading under illumination. Dark currents with high temporal noise components could render higher or lower EQEs than in reality. The mechanisms behind the increase of the EQE after the laser process remain unknown at the moment. Several dies have been sent to TEM to obtain more information. The improvement in the EQE over the non-implanted pixels could open the door for the optimisation of the devices using NLA processes in a new line of research.

The Ti implanted dies (with a dose of 3×10^{15} cm^{-2}) show a reduction in the EQE in all wavelengths, for almost all laser fluences studied. Short wavelengths are especially affected by the introduction of Ti atoms, as we could expect since the TIL is located in the surface, where the most energetic photons are absorbed. These results are consistent with what we showed in the previous chapter: photons with energy higher than the bandgap are converted to charges, but with lower efficiency than the non Ti implanted samples. The loss in quantum efficiency is very low, in any case, less than 5% over the total EQE in most wavelengths.

5.6.4 Individual Die Characterisation

Individual die characterisation allows to measure more parameters of the device, but at the increased cost of being more time consuming. In this section, we analyse several dies in chip carriers, with a Ti dose of 2×10^{15} cm^{-2}, laser annealed at different laser fluences and some non-implanted dies, with and without NLA process.

Since CVF results were shown in the WLC for the rest of the doses, we will omit this process here. The CVF values obtained for the dose of 2×10^{15} cm^{-2} varied between 71 and 75 μV/h$^+$. The next parameter of interest is the Full Well Capacity (FWC), which is the maximum amount of charges that can be retained inside the pixel. If any extra carrier is absorbed, it will not be able to be stored inside the potential well, so that it must be recombined elsewhere in the pixel (Fig. 5.23).

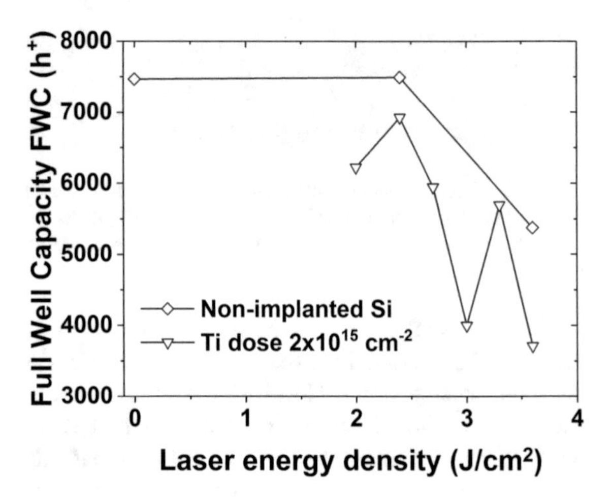

Fig. 5.23 Full well capacity of non-implanted and Ti implanted dies at a dose of 2×10^{15} cm^{-2} as a function of the laser fluence

The number of dies is limited, three non-implanted and six Ti implanted. We observe how the FWC does not change in the case of the non-implanted die after a laser process at 2.40 J/cm^2, but drastically decreases at the highest fluence (which corresponds to the range with broken pixels, Fig. 5.19). We observe a similar trend on Ti implanted dies, where in average, the FWC decreases with increasing fluence. At 2.4 J/cm^2, where pixels did not show any laser anneal defect (Fig. 5.19), we can compare the FWC of Ti implanted and non-implanted dies. The Ti implanted dies show a value 91% of the original FWC. We could attribute the decrease in the FWC to the volume occupied by the TIL. As measured in the previous chapter, we expect to find high Ti concentration values, which relate to high carrier concentration values. The potential well inside the pixel could be reduced due to the presence of the TIL. Assuming a layer 200 nm thick, where Ti concentration equals the concentration of the substrate (4×10^{15} cm^{-3}, see Fig. 5.9, it would correspond to the thickness of the metallurgical junction), and considering the thickness of the pixel, set to 3.5 μm, the TIL occupies around 6% of the total volume. Thus, the pixel would have 94% of the original volume for the collection of charges. The relative FWC measured on Ti implanted dies is close to this value and could be the main reason why the FWC diminished after the ion implantation process.

In Chapter 3, we fabricated pixel matrices. One of the main purposes was to check the homogeneity of the Ti supersaturated material. In Chapter 3, we found a high variability between neighbouring pixels, which we attributed to defects in the manufacturing route. Here, we have the possibility to examine the homogeneity of the material with large numbers of pixels (10^5 per family of pixels per die) and industrially standardised fabrication routes, through the measurement of the Photon-Response Non-Uniformity (PRNU). For this test, we measured the die annealed at 2.00 J/cm^2, which we found to show the worst results in terms on homogeneity (within the samples with laser fluences lower than 3.00 J/cm^2, see Fig. 5.19) PRNU as a function of the average hole level can be found on (Fig. 5.24).

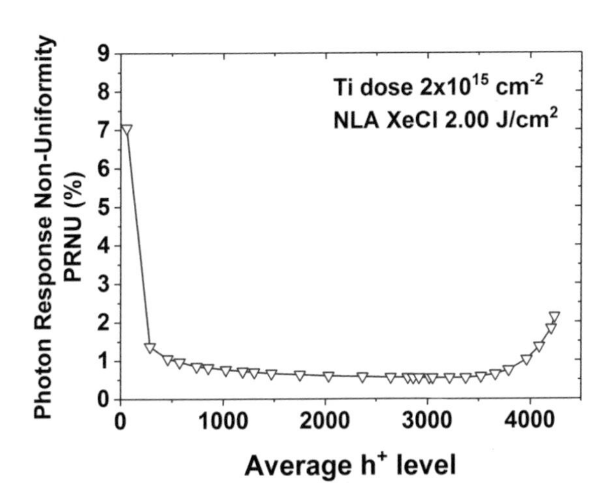

Fig. 5.24 Full well capacity as a function of the hole level for a die with a Ti dose of 2×10^{15} cm^{-2}, laser annealed at a fluence of 2.00 J/cm^2

PRNU is usually measured as a function of the average hole level stored inside the pixel, which is achieved by increasing the integration time (images with longer exposure). Then, a cell size is chosen to perform the averaging (10×10 pixels in this case). The average of each cell is compared and the standard deviation is extracted. There are three different regimes in the PRNU curve. The first one is placed at very low average hole levels. The PRNU asymptotically goes to infinitum at zero. This is related to the calculation of the standard deviation: when there are few holes inside the pixel, a small change is big in terms of relative variation. The extrapolation goes to the case when in some pixels there is one hole while in the other there are none. Then, the relative variation is infinite. When there are enough pixels (or cells) the value is stabilized (second region). In this sample, it is around 0.6–0.7%. The last region is characterized by a sudden increase in the PRNU, at the highest hole levels. This is mainly due to non-uniformities in the FWC. Some pixels may be already saturated while others are still being filled up with holes. PRNU level is usually compared between different technologies at half the FWC, which is around 0.62% in this case, a value really low. PRNU lower than 2% are usually considered sufficient to produce uniform pictures. Unfortunately, we could not measure the PRNU curve of a non-implanted die due to time restrictions, so we took the certified value at half the FWC of the non-implanted pixel structure instead. In our case, the reference dies exhibited values of 0.57%, which is not distant to the PRNU found on Ti implanted samples.

With this experiment, we could check that the introduction of both the Ti implantation and the NLA process did not affect the homogeneity of the material, which is another key aspect in this thesis. The integration of the Ti supersaturated Si layer resulted in homogenously responsive devices under illumination conditions, a necessary condition to obtain good images coming from the pixel matrix.

After the uniformity in illumination conditions, we evaluate the uniformity of the dark current, similarly to what we showed in Fig. 5.21. This is shown in Fig. 5.25.

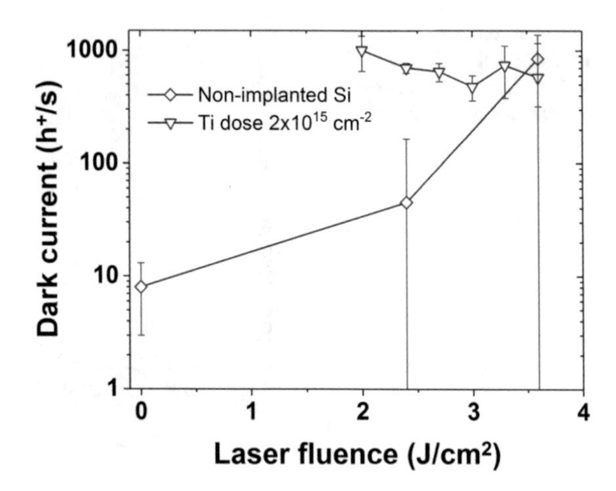

Fig. 5.25 Dark current as a function of the laser fluence for non-implanted and Ti implanted dies, at a dose of 2×10^{15} cm^{-2}

The results are similar to what was shown in Fig. 5.21: the laser process introduces a considerable amount of extra dark current. The Ti implanted samples exhibit dark current levels in the order of 400–900 h^+/s, close to what was obtained for the dose of 10^{15} cm^{-2} in Fig. 5.21. The uncertainty of the dark current is low at 2.40 and 2.70 J/cm^2, which points out to a homogeneous material, also in dark conditions.

Finally, we show the EQE curves. In individual dies it is possible to measure the whole QE curve, from 400 to 1000 nm, the limits of the experimental system. The measurement at longer wavelengths is still pending at CEA-LETI facilities, as several changes need to be done in the optical system to measure up to 1.6 μm.

The reference sample improves its EQE in the whole range after the NLA process of 2.40 J/cm^2, which is in line to what was shown in WLC in Fig. 5.22. Similarly, the EQE of the Ti implanted sample, annealed at the same laser fluence, show slightly lower EQE values, in the red and NIR part of the spectra. Its EQE is considerably reduced in the blue and green part of the spectra, as they are mostly absorbed near the TIL, which is expected to show higher recombination than the rest of the pixel area. This fact is further evidenced in Fig. 5.26 right, where we analyse the EQE at three different wavelengths as a function of the laser fluence. A behaviour as what was displayed in Fig. 5.22 is seen here: non-implanted but laser annealed dies exhibited higher EQE values than the reference, but Ti implanted showed slightly lower values.

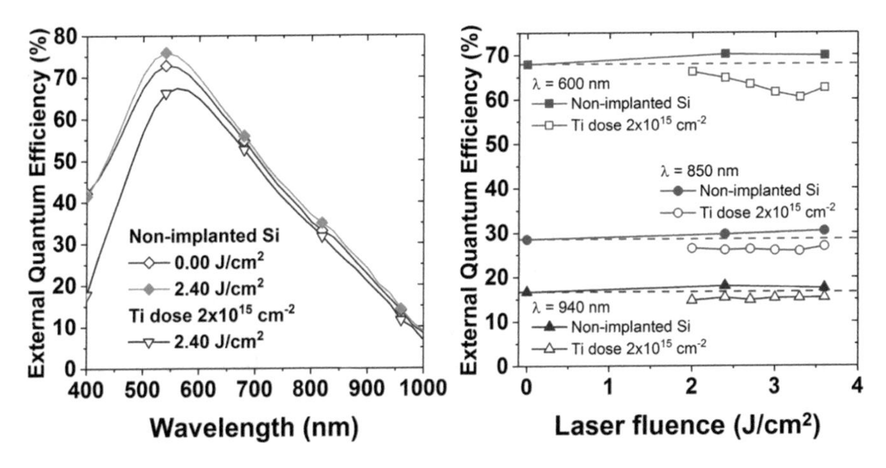

Fig. 5.26 External quantum efficiency as a function of the wavelength (left) for a non-implanted, non-annealed reference die, a non-implanted but laser annealed die and a Ti implanted and laser annealed die. Right: EQE as a function of the laser fluence for non-implanted and Ti implanted dies at a dose of 2×10^{15} cm^{-2} for several wavelengths

5.7 Conclusions

Along this chapter, we described (as far as possible, considering the NDA signed between our research group and STMicroelectronics) the integration, fabrication and measurement process of the Ti supersaturated Si material within a pre-commercial CMOS Image Sensor route.

In the first sections, we described the material fabrication and characterisation of the fabricated samples. Most of the characterisation was focused to find the appropriate laser fluences to obtain the best crystal quality possible, aiming to reduce the dark current and the inhomogeneities of the fabricated devices. Using several techniques, we identified four different recrystallization regimes, which were important to decide which laser fluences should be used in the patterned wafers, for final device fabrication. According to the material characterisation, laser fluences lower than 1.10 J/cm^2 did not melt the surface of the sample, not even the amorphised layer after the ion implantation process. Between 1.10 and 1.70 J/cm^2 the laser fluence was high enough to melt the amorphised layer, but was not high enough to melt the crystalline phase underneath it, so the resulting structure was polycrystalline. Increasing the fluence beyond 1.70 J/cm^2 up to around 2.40 J/cm^2 produced the best crystalline structures. Increasing further the laser fluence led to the apparition of a regrown amorphous layer, undesirable for the device performance.

The integration on the backside of the pixel was successful. Both the Ti ion implantation and the NLA processes were performed without any incompatibility issues. The levels of Ti contamination were within specifications in both CEA-LETI and STMicroelectronics facilities. The ROIC and the pixel structure kept their functionality and their integrity along the extra fabrication processes. The characterisation of the devices showed that the electrical parameters related to the ROIC were identical to those non-implanted with Ti. With respect to the pixel performance, we observed that the NLA process alone was responsible of most of the observed changes, the apparition of broken pixels at laser fluences higher than 3.00 J/cm^2 and an increase in the dark current, mainly. We measured that increasing the Ti dose led to lower dark current values, as opposed to what we measured in the previous chapter on transversal photodiodes fabricated in our facilities at UCM. The Ti implantation alone could be responsible of the decrease in the well capacity, which would be consistent with the volume physically occupied by the TIL within the pixel active area.

The measurement at die level allowed us to obtain the PRNU, which measures the uniformity of the pixel matrix in illumination conditions. The PRNU, which is the standard deviation over the average value of photoresponse, showed a non-uniformity of 0.62% at half the FWC, which is way below the recommended maximum PRNU of 2%. Under the same conditions, the reference die showed 0.57% of non-uniformity. The dark current measurements at die level pointed out to the same homogeneity in dark conditions, although the dark current was two orders or magnitude higher than the reference die. The homogeneity of the Ti supersaturated material is another key parameter for pixel matrix fabrication, required to certificate the functioning of any device at commercial level in the microelectronics industry.

Finally, the quantum efficiency, measured only for photon energies higher than the bandgap, showed improved EQE on non-implanted but laser annealed dies, which could open up a new field of research to further improve the performance of the fabricated pixels. The Ti implanted and laser annealed devices showed lower EQE in the same region, which was especially affected in the blue and green part of the spectra. The laser fluence, when kept between 2.40 and 2.70 J/cm^2 offered little variations in the EQE of the fabricated samples. Laser fluences lower than 2.40 J/cm^2 led to a sudden increase in the dark current, as it happened for fluences higher than 3.00 J/cm^2, where we observed that a considerable amount of pixels showed structural problems.

The development shown in this chapter leaves the door open for future collaborations with STMicroelectronics and CEA-LETI for further development of the technology, once the first proof-of-concept has been demonstrated. We expect to measure the fabricated dies in the sub-bandgap region, up to 1.6 μm at room temperature, with the collaboration of CEA-LETI, where they are installing a new experimental set-up in the near future.

References

1. Theuwissen A (2007) CMOS image sensors: state-of-the-art and future perspectives. In: Proceedings of the European Solid State Device Research Conference, pp. 21–27
2. Goiffon V et al (2014) Pixel level characterization of pinned photodiode and transfer gate physical parameters in CMOS image sensors. IEEE J Electron Devices Soc 2:65–76
3. Manouvrier JR, Fonteneau P, Montagner X (2015) Vertical gate transistor and pixel structure comprising such a transistor. US Patent US9209211B2

4. Wang M et al (2019) Thermal stability of Te-hyperdoped Si: atomic-scale correlation of the structural, electrical and optical properties. Phys Rev Mater 3: 044606
5. Olea J, Pastor D, Martil I, Gonzalez-Diaz G (2010) Thermal stability of intermediate band behavior in Ti implanted Si. Sol Energy Mater Sol Cells 94:1907–1911
6. Huet K, Mazzamuto F, Tabata T, Toque-Tresonne I, Mori Y (2017) Doping of semiconductor devices by Laser Thermal Annealing. Mater Sci Semicond Process 62:92–102
7. Hocine S, Mathiot D (1988) Titanium diffusion in silicon. Appl Phys Lett 53:3
8. Hocine S, Mathiot D (1989) Diffusion and solubility of titanium in silicon. Mater Sci Forum 38–41:725–728
9. Fisher DJ (1998) Diffusion in silicon: 10 years of research. Scitec Publications, Zuerich-Uetikon
10. Ishiyama K, Taga Y, Ichimiya A (1995) Reactive adsorption and diffusion of Ti on Si(001) by scanning tunneling microscopy. Phys Rev B 51: 2380
11. Auston DH, Surko CM, Venkatesan TNC, Slusher RE, Golovchenko JA (1978) Time-resolved reflectivity of ion-implanted silicon during laser annealing. Appl Phys Lett 33:437–440
12. Kerdiles S et al (2016) Dopant activation and crystal recovery in arsenic-implanted ultra-thin silicon-on-insulator structures using 308 nm nanosecond laser annealing. In: 2016 16th International Workshop on Junction Technology (IWJT), pp 72–75
13. Jellison GE, Burke HH (1986) The temperature-dependence of the refractive-index of silicon at elevated-temperatures at several laser wavelengths. J Appl Phys 60:841–843
14. Lowndes DH, Pennycook SJ, Jellison GE, Withrow SP, Mashburn DN (1987) Solidification of highly undercooled liquid silicon produced by pulsed laser melting of ion-implanted amorphous silicon: timeresolved and microstructural studies. J Mater Res 2:33
15. Donovan EP, Spaepen F, Turnbull D, Poate JM, Jacobson DC (1983) Heat of crystallization and melting-point of amorphous-silicon. Appl Phys Lett 42:698–700
16. Bonafos C, Mathiot D, Claverie A (1998) Ostwald ripening of end-of-range defects in silicon. J Appl Phys 83:3008
17. Asheghi M, Kurabayashi K, Kasnavi R, Goodson KE (2002) Thermal conduction in doped single-crystal silicon films. J Appl Phys 91:5079–5088
18. Fork DK, Anderson GB, Boyce JB, Johnson RI, Mei P (1996) Capillary waves in pulsed excimer laser crystallized amorphous silicon. Appl Phys Lett 68:2138–2140
19. Mathews J et al (2014) On the limits to Ti incorporation into Si using pulsed laser melting. Appl Phys Lett 104:112102
20. Liu F et al (2018) On the insulator-to-metal transition in titanium-implanted silicon. Sci Rep-UK 8:1–8
21. Liu XG et al (2014) Black silicon: fabrication methods, properties and solar energy applications. Energy & Environ Sci 7:3223–3263
22. Baumann AL et al (2012) Tailoring the absorption properties of Black Silicon. Energy Proced 27:480–484
23. Moon SJ, Lee M, Grigoropoulos CP (2002) Heat transfer and phase transformations in laser annealing of thin Si films. J Heat Transf 124:12
24. Miyasaka M, Stoemenos J (1999) Excimer laser annealing of amorphous and solid-phase-crystallized silicon films. J Appl Phys 86:5556–5565
25. Garcia-Hemme E (2015) Respuesta infrarroja en silicio mediante implantación iónica de metales de transición. Thesis dissertation
26. Theuwissen AJP (2008) CMOS image sensors: state-of-the-art. Solid State Electron 52:1401–1406

Chapter 6
Final Considerations

Along this chapter, we briefly describe the main conclusions derived from our research. This thesis started as the natural continuation of the thesis of E. García-Hemme [1], where he demonstrated that Ti supersaturated Si layers on macroscale devices, based on n-type substrates, could exhibit sub-bandgap photoresponse at room temperature, in the SWIR range, down to around 0.45 eV (3 μm).

As we discussed in Chapter 1, there is an increasing demand of SWIR photodetectors coming from the industry to satisfy the future needs of the society in diverse fields, as smart driving, Internet of Things, night vision or natural disaster management, among others. Most of the applications demand imaging sensors, which requires the use of FPA cameras featuring sensors in the microscale. With this idea in mind, we focused our research, within the frame of this thesis, towards the integration of the Ti supersaturated material into a microscale device, aiming to potentially integrate it into a commercial route in collaboration with a microelectronics company. The fabrication steps required to obtain Ti supersaturated Si layers are an ion implantation process followed by a NLA treatment to recover the crystal quality of the implanted layer. Both steps are commonly found in several fabrication routes in the microelectronics industry. Thus, the Ti supersaturated material has the potential to be integrated into a CMOS route, which could lead to low cost sensors, making the technology tentatively competitive with respect to other options.

Contributions of This Thesis

Since the Ti supersaturated Si layer already proved its performance in the macroscale, we designed a microscale prototype pixel structure, in the order of several hundreds of microns, along with the layout. The full process is described in Chapter 2. As a novelty in this thesis, we used p-type substrates in most of our research. The fabrication process included more than 100 different steps in our facilities at UCM, which were carefully characterised and optimised. After the fabrication process, we characterised the pixels by means of current-voltage and quantum

D. Montero Álvarez, *Near Infrared Detectors Based on Silicon Supersaturated with Transition Metals*, Springer Theses,
https://doi.org/10.1007/978-3-030-63826-9_6

efficiency measurements. We identified a dry etching problem that led to the discontinuity of the metallisation lines. Due to this problem, only 5% of the pixel worked, showing high variability between pixels of the same matrix. Among the functioning pixels, the results indicated that the TIL behaved as a n-type layer, as the pixels showed rectification at room temperature. The same pixels exhibited sub-bandgap photoresponse down to 0.45 eV (2.75 μm) at room temperatures, an energy considerably lower than the bandgap of Si. Besides, we measured that the photoresponse increased along with reverse polarisation, showing relative increments in the EQE of more than two orders of magnitude, although the polarisation increased the noise level as well. The best EQE value reported is 3.1% at the photon energy of 0.8 eV (1.55 μm), although cross-talk may be expected in our prototypes. The prototypes showed that Ti supersaturated devices in the microscale, based on p-type substrates, were technologically possible.

After the first prototypes showed promising results, we contacted STMicroelectronics (France) to develop together in collaboration a more mature prototype, in a pre-commercial CMOS route on p-type Si substrates. The collaboration led us to search for Ti implantation and NLA processes that could be compatible with the manufacturing lines of the company. Thus, we searched for a laser process that was certified for 300 mm wafers, as in the STMicroelectronics fab. We contacted a manufacturer of NLA instruments, SCREEN-LASSE, which provided access to the LT3100 tool. Since this laser equipment featured longer pulse durations (around 150 ns), when compared to the previous laser that we were using (25 ns), we fabricated a new set of samples aiming to characterise the new NLA process.

Chapter 4 describes the research done with the XeCl, long pulse duration laser. We swept the Ti dose and the laser fluence for each Ti dose, aiming to find the best range for sample fabrication with improved SWIR responsivity. We checked that most of the properties of the Ti implanted samples were similar to what was observed in our research group with the previous KrF laser. The characterisation process of the transversal, macroscale photodiodes allowed us to identify the different recrystallization regimes of the Ti implanted layers. We determined the minimum threshold energy that guaranteed a monocrystalline quality, with enough Ti concentration to potentially form the IB, and with sub-bandgap photoresponse. All the samples analysed in this chapter led to the photogeneration of carriers in the SWIR range, showing that the new laser was compatible with the fabrication of Ti supersaturated layers with sub-bandgap responsivity.

With the certainty that the laser equipment was compatible with both the formation of Ti supersaturated layers and 300 mm wafer handling capabilities, we started the collaboration project with STMicroelectronics. The project was carried out during a six-month internship in STMicroelectronics facilities in Crolles (France), which is described in Chapter 5. The project consisted in the integration of the Ti supersaturated material in a CMOS fabrication route designed by STMicroelectronics, including the fabrication and characterisation of prototype devices. We designed an integration plan of the Ti supersaturated material on 1 Mpx prototype CMOS sensors. The most important result is that both the Ti implantation and the NLA processes proved to be compatible with the CMOS route: no Ti cross-contamination was found during the fabrication process, the ROIC showed the same performance values as in reference samples and the Ti implanted pixels were light-responsive with similar performance. Besides, the uniformity of the Ti supersaturated pixels was inside specifications, very close to the reference pixels, which further certifies that the supersaturated material is homogeneous under industry standards. We measured that most of the effects seen in the fabricated devices were due to the laser process, namely higher dark currents and loss of pixel integrity at very high laser fluences (above 3.00 J/cm^2). From the analysis of the results, we observed that non-implanted but laser annealed pixels improved the EQE values in the whole curve, from 300 to 1000 nm, which may open a new line of research aiming to further increase the EQE in the visible and NIR range. Ti implanted dies exhibited performance values close to the reference, with slightly lower EQE values in the same range. The measurement of the EQE in the sub-bandgap region is still pending, as the optical characterisation benches available at Crolles during the internship were not calibrated for wavelengths longer than 1 μm.

Thus, the main goals originally proposed in this thesis have been fulfilled: the Ti supersaturated Si material developed in our research group, with sub-bandgap absorption in the SWIR range at room temperature, was successfully integrated in a pre-commercial Si CMOS route, establishing a fluid collaboration with international organisms, from both the industry (STMicroelectronics, SCREEN-LASSE and Ion Beam Services) and public research centres as CEA-LETI-Minatech. The collaboration has been fruitful and continues to date to further develop the technology derived from this thesis.

Reference

1. Garcia-Hemme E (2015) Respuesta infrarroja en silicio mediante implantación iónica de metales de transición. Thesis dissertation

Chapter 7
Future Progress

The results derived from this thesis (briefly discussed in the previous chapter) leave the door open for further development and optimisation in several branches:

- The findings of Chapter 3 may lead to the seventh and newer generations of Silfrared devices, developed and manufactured at UCM. The research level at the university level is also crucial as the fabrication process of complete wafer lots at STMicroelectronics exceeds one year from wafer fabrication to device characterisation. In our laboratories, the devices can be measured in less than one month after the wafers are implanted.
- During this thesis, the multi-cathode sputtering and the wire bonding equipment were installed and optimised. There is still room for improvement in our technology, especially with the sputtering system, which is still under development and must be finished to provide the best metallisation results. From our measurements in Chapter 3, we concluded that metallic contact deposited by sputtering provided ohmic contacts with improved properties, as compared to e-beam evaporation.
- In this thesis, only p-type substrates were analysed using microscale devices. It would be interesting to complete the characterisation of the n-type substrates in the form of pixel matrices.
- In Chapter 4, we used a four-layer model that fitted the T-R measurements. It allowed the estimation of the complex refractive index of the TIL. We suggest to further implement the model with new simulators in order to improve the results.
- The conduction properties of TIL over p-type Si substrates were studied for the first time using the van der Pauw configuration. This led to the development a tri-layer model that could explain the results. This tri-layer model needs further developing, aiming to obtain more information on the conduction processes of the TIL.
- The effect of the XeCl laser of n-type substrate is not explored to the same extent as in p-type substrates. We propose to fabricate Ti supersaturated devices using n-type substrates.

D. Montero Álvarez, *Near Infrared Detectors Based on Silicon Supersaturated with Transition Metals*, Springer Theses,
https://doi.org/10.1007/978-3-030-63826-9_7

- Given the importance of the laser process in the properties of the Ti supersaturated material, it would be interesting to acquire or develop a simulation program to predict the effect of different laser processes (pulse duration, wavelength) on different Si substrates (crystalline or amorphised with different thicknesses).
- Most of the measurements were performed at room temperature. We propose to analyse the current-voltage characteristics and the EQE at variable temperature, which may help in determining the origin of the conduction processes and the mechanisms producing the high reverse current.
- After comparing samples annealed with the same NLA equipment but with different Ti implantation processes, mainly the ion implantation energy, we propose to perform a study at a given Ti dose, sweeping the energy of the implantation process. In this thesis, we observed that 150 keV was too high as most of the laser processes did not melt the whole implanted layer, while 10 keV might have been too low, due to the presence of a regrown amorphous layer at the surface.
- In Chapter 5, we chose a pixel structure, where we integrated the Ti supersaturated material. We used four Ti doses and several laser annealing processes. However, mainly due to time restrictions, we could only fully analyse the samples with the Ti dose of 2×10^{15} cm^{-2}. We propose to continue the analysis of the rest of the Ti doses in collaboration with STMicroelectronics. Besides, EQE in the sub-bandgap region was not measured. The collaboration, including CEA-LETI is expected to continue mounting a new optical bench capable to go down to, at least, 0.77 eV (1.6 μm). The sub-bandgap photoresponse of the CMOS devices is still unexplored and could be the key to further develop the industrial-based prototype, towards a future commercialisation of Ti supersaturated devices for sensing in the SWIR range.
- In line with the collaboration with STMicroelectronics, we propose to use different Ti implantation energies to solve the problem of the regrown amorphous layer.

The collaboration with STMicroelectronics has allowed me to check first-hand the big differences, with its advantages and disadvantages, of the research performed at university and at a private company level. I believe that combining the best of both worlds could be beneficial for further development of the technology, hence the importance of continuing the collaboration between the University and STMicroelectronics.

At the present moment, the collaboration with STMicroelectronics and CEA-LETI is ongoing. We have send an original manuscript describing the recrystallization processes of Ti implanted samples for the peer-review process in a journal within the first quartile, and we have a second manuscript undergoing the writing and revising process. There are programmed meetings in the next months to decide the future of the project, involving the possibility to apply for a position at STMicroelectronics, aiming to continue the project on Ti supersaturated CMOS Image Sensors. Our research group at UCM has recently been granted with funding for the second stage of the project of supersaturated materials, which further evidences the interest of the public funding organisms on the project. The funding also guarantees that the research will continue for at least three years more.

Appendix

The information contained in this chapter is not strictly necessary to understand the core of the thesis, but it will provide further understanding in those fields that were not completely explained in the text due to space restrictions.

Appendix A: Previous Research

This annex contains a brief summary of the material properties previously published in our research group related to Ti supersaturated Si samples, using a double ion implantation process, with a total dose of 5×10^{15} cm^{-2} and laser annealed using the KrF laser described in Sect. 2.1.4. This section is a brief summary that may be useful for those readers not familiar with supersaturated materials. For further information, please refer to the cited references along this annex.

Ti Profile Distribution: SIMS Technique

The supersaturating condition is achieved when a certain impurity concentration level is achieved, the IB formation limit. We use ToF-SIMS technique to obtain the Ti profile distribution of an as-implanted and a laser annealed sample at 1.8 J/cm^2, both implanted with the same Ti dose. The as-implanted profile exhibits a double peak feature, which is due to the double ion implantation process:

Figure A.1 represents the Ti concentration as a function of the depth for two different Ti implanted samples: one as-implanted sample and another with a NLA process using a laser fluence of 1.8 J/cm^2. The as-implanted curve shows an almost constant Ti profile up to around 170 nm, where it falls down. The curve of the laser annealed sample points out to a Ti redistribution, where atoms have been

D. Montero Álvarez, *Near Infrared Detectors Based on Silicon Supersaturated with Transition Metals*, Springer Theses,
https://doi.org/10.1007/978-3-030-63826-9

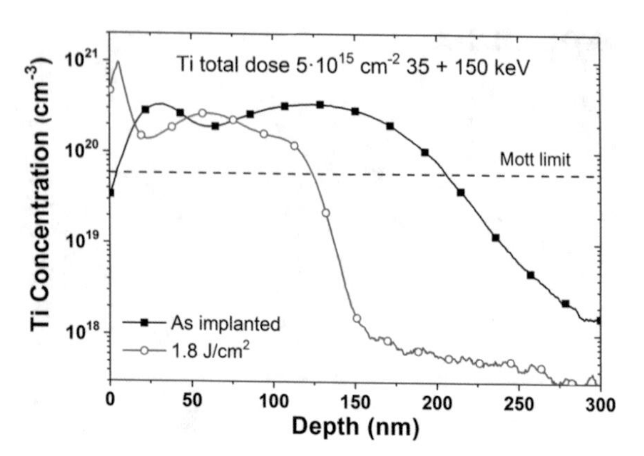

Fig. A.1 ToF-SIMS Ti distribution profile of a sample with a double ion implantation process with a Ti dose of 5×10^{15} cm^{-2}. Curves for the as-implanted and annealed at 1.8 J/cm^2 samples are shown. Original data extracted from Olea et al. [1]

pushed towards the surface, due to the snow-plow effect [2]. In this curve, the peak concentration is found at around 10 nm, with a concentration peak of 10^{21} cm^{-3}. The concentration drops down to around 1–2 $\times$ 10^{20} cm^{-3}, where it is more or less constant up to 125 nm, where it drops down to a concentration level of 10^{18} cm^{-3} up to the noise level, around 3×10^{17} cm^{-3}.

It is possible, using the as-implanted curve as reference, to estimate the remaining dose of Ti inside the sample after the NLA process. The area under the as-implanted curve is equal to the dose, 5×10^{15} cm^{-2}. If we perform the same calculus under the curve of the sample annealed with 1.8 J/cm^2, we find a retained dose of 2.86 $\times$ 10^{15} cm^{-2}, which is around 57.2% of retained dose. We suspect it may be due to Ti migration outside the wafer, towards the atmosphere, during the melting and subsequent solidification process [3].

According to this Ti profile, the wafers used for the fabrication of pixel matrices would have a supersaturated layer around 125 nm thick, while the total implanted layer would be around 320 nm thick, considering the noise level at 3×10^{17} cm^{-3}. The next step is to evaluate the crystal quality of the implanted layer.

Crystal Quality: XTEM Micrographs

One of the most common techniques to check the crystal quality is by means of Cross-Sectional TEM (XTEM). The results described here are based on the work of Olea et al. [1].

Figure A.2 displays the XTEM micrograph of a sample with the same fabrication parameters as the ones fabricated in this subsection of the thesis. In overall, the quality of the implanted layer is very good, argument supported by the Electron Diffraction (ED) patterns insets shown on the right. There is almost no difference between the two ED patterns; the upper one corresponds to the implanted layer, while the lower one corresponds to the area close to the substrate. There are still some defects visible close to the surface, with an average feature size of 119 nm, value that is close to the thickness of the supersaturated layer. These features have been documented in the literature, and they have been labelled "cellular breakdown"

Fig. A.2 XTEM micrograph of a sample double ion implanted with a total Ti dose of 5×10^{15} cm^{-2}, subsequently laser annealed with a fluence of 1.8 J/cm^{2}. Original TEM micrographs from Olea et al. [1]

structures [4]. Those structures are filamentary-like columnar structures which are rich in Ti atoms. However, their influence on the electrical and optical properties of the supersaturated material are not yet fully understood [4–6].

Thus, the NLA process seems suitable to produce monocrystalline Ti supersaturated Si layers.

Electro-Optical Properties: T-R and Responsivity Measurements

The Ti supersaturated material has been shown to provide good crystal quality while having enough Ti concentration to possibly form an impurity band. This sub-section shows the electro-optical properties of the supersaturated material.

Transmittance and Reflectance (T-R) measurements shown here are taken from Olea et al. [1], while responsivity measurements are taken from García-Hemme et al. [7]. First, T-R are measured in a reference wafer, which is non-implanted. Then, the Ti implanted wafer is measured.

From Fig. A.3 we see that the reflectance of the Ti supersaturated sample is lower than the reference, except for the as-implanted sample at values close to the bandgap. Together with the transmittance values, which are also lower than the reference sample, we can conclude that the absortance of the implanted samples is higher than that of the reference in the range under study, from 0.55 to 1.2 eV. There is a big difference between the as-implanted and the laser annealed sample. The as-implanted sample exhibits what may look like periodic oscillations, possibly due to reflections between the amorphous layer, formed during the ion implantation process, and the substrate. The possible mechanism that could explain the increase in the absortance were discussed in the manuscript, where it was inferred that the

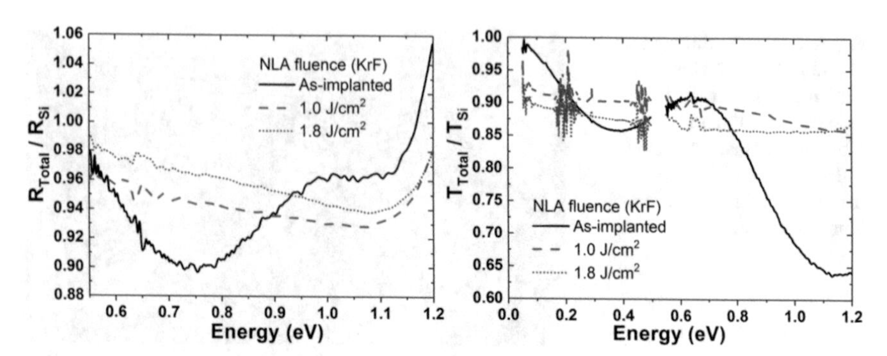

Fig. A.3 Reflectance (left) and transmittance (right) of the as-implanted and the NLA processed sample at 1.8 J/cm^2. Ti total dose is 5×10^{15} cm^{-2}. Original data extracted from Olea et al. [1]

most plausible explanation for the sub-bandgap absorption could be the formation of an Impurity Band, mediated by Ti atoms.

The supersaturated layer has been proved to be absorbent in the sub-bandgap regime, but, in a good photodetector, not only the light needs to be absorbed, but it needs to be converted into a measurable electrical signal. In order to have a measurable electrical signal (current or voltage) coming from an optical excitation, the generation rates must be higher than the recombination rates within the semiconductor. We use different techniques to measure the photogenerated current or voltage, as explained in Sect. 2.4.5. To characterise the material described in this sub-section, we rely on the work of García-Hemme et al. [7], where our research group measured the spectral photovoltage measured under the van der Pauw configuration of a 1 cm^2 sample. The voltage was used to calculate the responsivity of the sample, which is shown below, at room temperature:

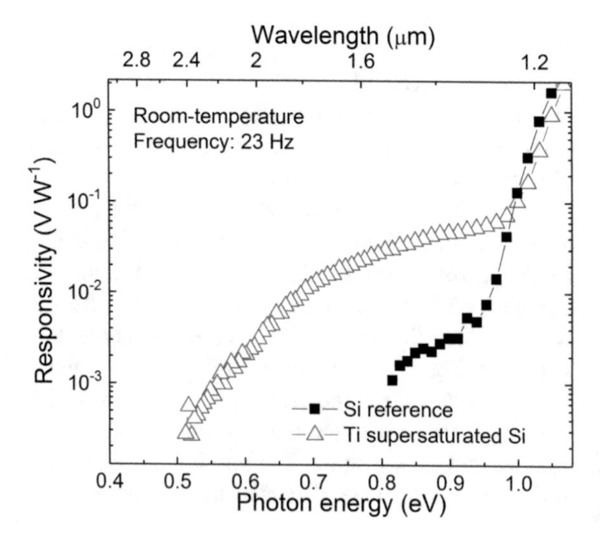

Fig. A.4 Responsivity of a double ion implanted sample with a total Ti dose of 5×10^{15} cm^{-2}, subsequently followed by a NLA process with a fluence of 1.8 J/cm^2. Measurement done at room temperature. Original data extracted from Garcia-Hemme et al. [7]

The responsivity displayed in Fig. A.4 shows that the supersaturated material may be suitable to fabricate detectors in the NIR and SWIR part of the spectra, at room temperature, without the use of any cooling system. The same figure shows the responsivity of a non-implanted Si reference sample. The reference sample exhibits an abrupt decay in its responsivity at around 1.1 eV. There is some sub-bandgap response on the reference sample, which rapidly reaches the noise level, which could be associated to surface defects [7, 8]. The implanted sample, however, shows a similar drop in responsivity for energy values close to the bandgap, although the responsivity in the sub-bandgap region (energies lower than 1.1 eV) is up to two orders of magnitude higher than the reference. In fact, not only its responsivity is measurable up to 0.5 eV, but its noise level is almost one order of magnitude lower than the reference as well. In the referred work, several possibilities are considered to explain the sub-bandgap photoresponse. Among the different options, the photogeneration due to an impurity band caused by Ti atoms is suggested over the other possibilities, as Free Carrier Absorption (FCA) or two photon absorption processes.

Ruling Out the Impact of Bulk Defects on the Sub-bandgap Photoresponse of Ti Implanted Si Layers

This sub-section briefly describes the experiments from Olea et al. [9] where they studied the effect of bulk defects by performing a self-implantation on Si substrates.

The substrate used in this work is the same used in Sect. 3.3. The Si PAI process is slightly different, with a higher dose of 10^{16} cm^{-2} and 170 keV. The measurements were performed at 90 K to reduce the noise found at room temperature in the samples. These measurements seem to rule out the possible origin of sub-bandgap photoresponse in the damage caused by the ion implantation process, nor the NLA process. As seen in Fig. A.5, self-implanted samples, subsequently annealed, did not show any significant improvement on the sub-bandgap photoresponse as compared

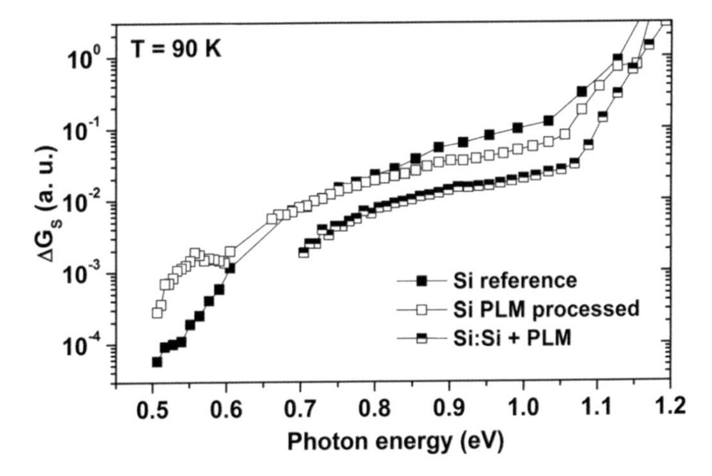

Fig. A.5 Sub-bandgap photoconductance as a function of the photon energy for three different Si samples, one non-implanted and non-annealed, another non-implanted but laser annealed, and a third self-implanted, followed by a PLM process at 90 K. Original data taken from Olea et al. [9]

to the reference. To date, the presence of Ti in high concentrations seems to be the only plausible explanation for the sub-bandgap photoresponse found in the Ti supersaturated samples.

Appendix B: List of Fabrication Routes

This chapter contains the different fabrication routes (called generations) used to manufacture the pixel prototypes developed at UCM in the frame of this thesis. Each generation showed incremental improvements until the last generation, the sixth.

First Generation

The first generation was used to test the different fabrication steps and to familiarise ourselves with the microelectronic techniques. Thus, the pixel is simplified to a photoconductive pixel with no mesa structure (Table B.1).

We detect that the metallisation process is easily removed after ultrasound cleaning processes. The adherence problem limited the measurements of IV curves as the

Table B.1 Fabrication route of Silfrared devices of first generation

First generation			
Process	Wafer	SR1.1	SR1.2
Cleaning	First cleaning	N_2 blow	
Deposit	SiO_2	150 nm	
Wafer preparation for lithography	Pre-bake	145 °C 30 min	
	Cooling	6000 rpm 30 s twice	
	Depo TI Primer	6000 rpm 30 s	
	Depo PR AZ 4533	4000 rpm 30 s	
	Soft-bake	100 °C 3 min	
	Cooling	30 min	
Positive lithography: "Pad opening"	Alignment	Flat, centred motifs	
	Exposition	22 s, Mode I1, timer, control	
Development and stop bath	Development AZ 826 MIF	20 mL 3 min	
	Stop bath	DI water	
	Hard-bake	135 °C 5 min	
	Cooling	Until 80 °C on surface	
SiO_2 etching	BOE etching	20 mL 3 min	

(continued)

Table B.1 (continued)

First generation		
	Stop bath	DI water
Photoresist etching	Etching DMSO	20 mL 4 min
	Stop bath	DI water 2 min
Deposition	Aluminium	200 nm
Wafer preparation for lithography	Pre-bake	145 °C 30 min
	Cooling	6000 rpm 30 s twice
	TI Primer	6000 rpm 30 s
	Photoresist AZ 4533	4000 rpm 30 s
	Soft-bake	100 °C 3 min
	Cooling	30 min
Positive lithography: metallisation	Alignment	Pad opening motifs
	Exposition	22 s, Modes I1, timer, control
Development and stop bath	Development AZ 826 MIF	20 mL 3 min
	Stop bath	DI water
	Hard-bake	135 °C 5 min
	Cooling	Until 80 °C
Al etching	Etching NaOH	1 g NaOH + 100 mL H_2O 2 min
	Stop bath	Agua DI
Photoresist etching	Etching with DMSO	20 mL 4 min
	Stop bath	DI water 2 min twice

probes could break the metallic lines. Wire bonding to these structures was not possible.

Second Generation

This set of samples is intended to improve the metallisation.

Table B.2 Fabrication route of Silfrared devices of second generation

Second generation					
Process	Wafer	SR2.1	SR2.2	SR2.3	SR2.4
Cleaning	First cleaning	N_2 blow			
Deposition	SiO_2	150 nm			
Wafer preparation for lithography	Pre-bake	145 °C 30 min			
	Cooling	6000 rpm 30 s twice			
	TI Primer	6000 rpm 30 s			
	Photoresist AZ 4533	4000 rpm 30 s			
	Soft-bake	100 °C 3 min			
	Cooling	30 min			
Positive lithography: pad opening	Alignment	Flat, centred motifs			
	Exposition	22 s, Modes I1, timer, control			
Development and stop bath	Development AZ 826 MIF	20 mL 3 min			
	Stop bath	DI water twice			
	Hard-bake	135 °C 5 min			
	Cooling	Until 80 °C			
Etching SiO_2	Etching BOE	20 mL 3 min			
	Stop bath	DI water twice			
Photoresist etching	Etching DMSO	20 mL 4 min			
	Stop bath	DI water 2 min twice			
Deposition	Aluminium	200 nm			
	Gold	250 nm			
Wafer preparation for lithography	Pre-bake	145 °C 30 min			
	Cooling	6000 rpm 30 s twice			
	TI Primer	6000 rpm 30 s			
	Photoresist AZ 4533	4000 rpm 30 s			
	Soft-bake	100 °C 3 min			
	Cooling	30 min			
Positive lithography: metallisation	Alignment	Pad opening motifs			
	Exposition	22 s, Modes I1, timer, control			
Development and stop bath	Development AZ 826 MIF	20 mL 3 min			
	Stop bath	DI water			
	Hard-bake	135 °C 5 min			
	Cooling	Until 80 °C			
Metal etching	Gold etching	4 g KI + 1 g I_2 + 40 ml H_2O 1 min			
	Stop bath	DI water twice			
	Aluminium etching	1 g NaOH + 100 mL H_2O 2 min			
	Stop bath	DI water twice			

(continued)

Table B.2 (continued)

Second generation		
Photoresist etching	Etching with DMSO	20 mL 4 min
	Stop bath	DI water 2 min twice

We proposed the use of FGA processes to see if we could improve the rectifying contacts observed in the first generation. In order to improve the wire bonding, we include in the recipe the deposition of Au layers. We also introduce an extra cleanliness step on all wet etching process.

The introduction of gold in the route led to the introduction of an extra step of Au etching. We found problems in the gold etching process, which resulted in poor wire bonding processes, gold was not correctly adhered to the Al layer. We attribute the problem to a possible incompatibility between the DMSO and DI water. We propose to change the stop bath of this step to IPA instead. After FGA at 700 °C the devices show similar adherence problems.

Third Generation

In the third generation, we reduce the temperature of the FGA processes to 500 °C. To avoid repetition, in next Table B.3 we only include the new or modified steps.

Table B.3 Fabrication route of Silfrared devices of third generation

Third generation				
Process	Wafer	SR3.1	SR3.2	SR3.3
Cleaning	First cleaning	N2 blow		
Deposition	SiO_2	150 nm		
Wafer preparation for lithography	Go to Table B.2			
Positive lithography: pad opening				
Development and stop bath				
Etching SiO_2				
Photoresist etching	Etching with DMSO	20 mL 4 min		
	Stop bath	IPA 2 min twice		
Deposition	Aluminium	200 nm		

(continued)

Table B.3 (continued)

Third generation				
FGA	Al doping 500 °C	10 min	5 min	3 min
Deposition	Gold	NO		250 nm
Wafer preparation for lithography	Go to Table B.2			
Positive lithography: metallisation				
Development and stop bath				
Metal Etching				
Photoresist etching	Etching DMSO	20 mL 4 min		
	Stop bath	IPA 2 min twice		

The problems with the Au layer continued, even after FGA with lower temperatures. We decided to go back to Al layers only, as Au only increased the number of steps.

Fourth Generation

We incorporated for the first time Ti implanted dies in this generation on the backside of the wafer. We performed a preliminary study where we obtained the best adherence results of Al layers with shorter FGA processes (Table B.4).

Table B.4 Fabrication route of Silfrared devices of fourth generation

Fourth generation				
Process	Wafer	SR4.1	SR4.2	SR4.3
Ion implantation	Titanium	NO	NO	YES
NLA	Laser KrF 25 ns	NO	NO	YES
Cleaning	Go to Table B.2			
SiO_2 Deposition				
Wafer preparation for lithography				
Positive lithography: pad opening				
Development and stop bath				
Etching SiO_2				
Photoresist etching				
Deposition	Aluminium	200 nm		
FGA	Al doping 500 °C	1 min	2 min	2 min
Wafer preparation for lithography	Go to Table B.2			

(continued)

Table B.4 (continued)

Fourth generation		
Positive lithography: metallisation		
Development and stop bath		
Etching metal		
Photoresist etching		
Wafer cleaving	Diamond point	4 fragments

The wire-bonding process shows improved results with respect to the previous generations. We measure the first sub-bandgap EQE in the Ti implanted dies. However, for energies close to the bandgap we measure EQE higher than 100%, which leads to crosstalk.

Fifth Generation

In order to evaluate the proper EQE we designed an extra mask that could cover several pixels. It requires an extra lithography step.

Table B.5 Fabrication route of Silfrared devices of fifth generation

Fifth generation						
Process	Wafer	SR5.1	SR5.2	SR5.3	SR5.4	SR5.5
Ion Implantation	Titanium	YES	YES	NO	NO	NO
Wafer cleaving	Diamond point	4 fragments				
NLA	Laser KrF 25 ns	YES	YES	NO	YES	NO
Cleaning	Go to Table B.2					
Deposition						
Wafer preparation for lithography						
Positive lithography: pad opening						
Development and stop bath						
Etching SiO_2						
Photoresist etching						
Deposition	Aluminium	200 nm				
FGA	Al 500 °C	2 min				
Wafer preparation for lithography	Go to Table B.2 Go to Table B.2					

(continued)

Table B.5 (continued)

Fifth generation						
Positive lithography: metallisation						
Development and stop bath						
Etching metal						
Photoresist etching						
Wafer preparation for lithography	Pre-bake	145 °C 30 min				
	Cooling	6000 rpm 30 s twice				
	TI Primer	6000 rpm 30 s				
	Photoresist nLof AZ 2035	3000 rpm 30 s				
	Soft-bake	110 °C 1 min				
	Cooling	30 min				
Negative lithography: covering pixels	Alignment	Metallisation motifs				
	Exposition	20 s, I2, timer, control				
	Post-bake	100 °C 2 min				
	Cooling	Until 80 °C				
Development and stop bath	Development AZ 826 MIF	20 mL 3 min				
	Stop bath	DI water				
Deposition	SiO_2 + Al + SiO_2	NO	YES	NO	NO	YES
Lift-off	Etching DMSO	20 mL 1 hour 80 °C + ultrasounds				

Few pixels were measured, the lift-off step was not correct and we could wire bond only a few pixels of the non-implanted wafers. Covering the pixels is abandoned until lift-off is better controlled for this step.

Sixth Generation

We finally arrive the most mature generation, which is profusely described in Sect. 3.5. The main difference is that we include the mesa structures, which implies an extra lithography process. Besides, we change the positive lithography process of the metallisation process for a negative lithography process prior to Al deposition. Later, we use a lift-off process to remove the excess Al. The number of samples is increased, as six wafers were measured, one of each with up to four dies. The lithography processes did not change with respect to what was described in Table B.5, so it is not included here.

Appendix C: The Four-Layer Model

This annex describes the four-layer model used to fit the experimental T-R data of Ti implanted samples (Sect. 4.3.3). The equations are taken from Maley et al. [10].

First, we define the partial reflection and transmission coefficients, only valid for adjacent layers, that is, $j = i \pm 1$:

$$t_{ij} = \frac{2N_i}{N_i + N_j} \tag{C.1}$$

$$r_{ij} = \frac{N_i - N_j}{N_i + N_j} \tag{C.2}$$

Later, these values are used to calculate the individual reflection and transmission coefficients:

$$R_{ij} = \left| r_{ij} \right|^2 \tag{C.3}$$

$$T_{ij} = \frac{n_j}{n_i} \left| t_{ij} \right|^2 \tag{C.4}$$

$$P = e^{-\frac{2\pi i N_2 d_{TIL}}{\lambda}} \tag{C.5}$$

Where n in lowercase stands for the real part of the complex refractive index, d_{TIL} is the thickness of the Ti implanted layer and λ is the wavelength of the light. The next step is to calculate the reflectance and the transmittance between layers 1 and 3, which are equivalent to assume that the substrate is semi-infinite:

$$T_{13} = T_{31} = \frac{n_3}{n_1} \left| \frac{t_{12} t_{23} P}{1 - P^2 r_{21} r_{23}} \right|^2 \tag{C.6}$$

$$R_{13} = \left| r_{12} + \frac{P^2 t_{12} t_{21} r_{23}}{1 - P^2 r_{21} r_{23}} \right|^2 \tag{C.7}$$

$$R_{31} = \left| r_{32} + \frac{P^2 t_{32} t_{23} r_{21}}{1 - P^2 r_{21} r_{23}} \right|^2 \tag{C.8}$$

Which are used to finally calculate the total transmittance and reflectance:

$$T = T_{14} = \frac{T_{13} T_{34} e^{-\alpha d}}{1 - R_{43} R_{31} e^{-2\alpha d}} \tag{C.9}$$

$$R = R_{14} = R_{13} + \frac{T_{13} T_{31} R_{34} e^{-2\alpha d}}{1 - R_{34} R_{31} e^{-2\alpha d}} \tag{C.10}$$

In the last two equations, d refers to the thickness of the wafer, instead of the TIL. The model uses as inputs the real and imaginary parts of the complex refractive index, n and k respectively. We model the real part n following a Cauchy dispersion relation:

$$n_2(\lambda) = A + \frac{B}{\lambda^2} \tag{C.11}$$

Which may be suitable up to the refractive index peak found on bare Si, at around 380 nm. With respect to the imaginary part of the refractive index, k_2, we took a three-order polynomial approximation, as seen in other works [11]:

$$k_2(\lambda) = ax^3 + bx^2 + cx + d \tag{C.12}$$

We imposed an extra boundary condition, to simulate the absorption coefficient of the samples in the UV region, which is expected to be similar to that of crystalline silicon. We choose the maximum value between the k_2 value or the k_3 value of silicon:

$$k_2(\lambda) = \max\left(ax^3 + bx^2 + cx + d, k_3(\lambda)\right) \tag{C.13}$$

Finally, we obtain the absorption coefficient through the imaginary part of the complex refractive index as:

$$\alpha = \frac{4\pi k}{\lambda} \tag{C.14}$$

Appendix D: The Conductance Bi-Layer Model

This annex contains a brief description of the bilayer model developed by Olea et al. [12]. For further reading, please refer to the cited manuscript.

The horizontal electrical measurements on Ti implanted samples used in this thesis can be described as two layers connected in parallel: the surficial one being the Ti implanted layer and the second one being the Si substrate. Assuming that the lateral dimension l of the sample is much bigger than the sum of the thicknesses of both layers t, i.e. $l \gg t$, the current can be assumed to flow parallel to the surface, as depicted in Fig. D.1:

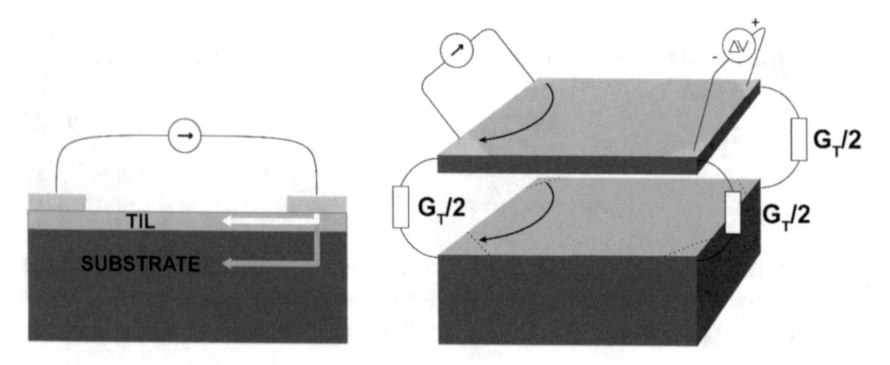

Fig. D.1 Cross-section view of the bilayer model, indicating the expected current flow direction assuming a thin sample

The four contacts are deposited in the corners on the surface, that is, in direct contact with the Ti implanted layer. Depending on the electrical nature of the junction between the TIL and the substrate, there could be an electrical barrier that could modify the current in-depth distribution, possibly preventing the current to flow through the substrate. In previous works in our research group [12] this current blocking mechanism has been modelled using a transversal conductance $G_T/2$ under each contact. The full bilayer model is explained in the previous reference. We will use the conductance, instead of the resistance, as it is easier to operate with conductances in parallel. The total conductance of two layers in parallel can be written as follows:

$$G_{SH} = \frac{(G_{TIL}+G_{SUB}F)^2}{G_{TIL}+G_{SUB}F^2} \tag{D.1}$$

Where G_{SH} is the total conductance of both layers, G_{TIL} is the conductance of the TIL, G_{SUB} is the conductance of the substrate and F is what has been called the "decoupling function" [12], defined as:

$$F(T) = \frac{G_T(T)}{G_T(T)+\frac{G_{SUB}(T)}{\alpha}} \tag{D.2}$$

The decoupling function is defined in the [0,1] interval. When the transversal conductance is zero, then $F = 0$, which means that the substrate is not participating in the conduction processes (it is electrically isolated). Under these circumstances, the TIL is said to be decoupled from the substrate. On the other side, if the transversal conductance is infinite, then the substrate and the TIL are electrically short-circuited, implying that the function $F = 1$. In that case, the TIL and the substrate are said to be electrically coupled. If the transversal conductance has a finite value higher than zero, then the value of the decoupling function is lower than the unity. The value of the decoupling factor may vary with temperature, as the transversal and the substrate conductances may also vary with temperature at different rates. In Eq. 4.7, G_T is the previously defined transversal conductance and α is a geometrical parameter, defined as the constant of proportionality between the resistance between two neighbouring contacts and the sheet resistance in the van der Pauw configuration. The value of α is related to the geometry of the sample and is assumed to be equal to 2 for the sample and contact size used in this thesis, as described in our previous works [12, 13]. The value of G_T is temperature dependent, and it was demonstrated to follow an Arrhenius law, as previously published in our group by Olea et al. [12]:

$$G_T(T) = G_0 e^{-\frac{\Delta E}{kT}} \tag{D.3}$$

Garcia-Hemme et al. [14] continued and reinterpreted the work by Olea et al. In the previous equation, G_0 is the transversal conductance at 0 K, ΔE has been interpreted as the height of the electrical barrier between the TIL and the substrate, k is the Boltzmann constant and T is the temperature. In the same work by Garcia-Hemme et al., they applied the Meyer-Neldel rule to the supersaturated Si layers, which relates the pre-exponential factor and the activation energy as:

$$G_0 = G_{00} e^{-\frac{\Delta E}{kT_{MN}}} \tag{D.4}$$

Where G_{00} is the pre-exponential factor and T_{MN} is the Meyer-Neldel temperature. The Meyer-Neldel rule applies to different samples having the same blocking mechanism, in our particular case different Ti implanted samples, with either different Ti doses or NLA processes. The model predicts the same transversal conductance G_{00} at the same temperature T_{MN} for all the samples that follows the rule. The nature of the blocking mechanism on supersaturated samples has been linked to the height of the electrical barrier caused by the implantation tails, present at the interface between the TIL and the substrate [14]. The ion implantation tails are implanted atoms that are not in concentrations high enough to produce an impurity band (in our case, Ti atoms), which could act as recombination centres at the interface between the TIL and the substrate.

The electrical characterisation continues by applying the bilayer model to obtain the mobility and the carrier concentration of each layer. The experimental measurements will provide effective values, which would take into account the conduction mechanisms of both layers. The equations to relate the effective Hall mobility, μ_{eff} and the effective sheet carrier concentration, n_{eff}, which were defined in Eq. 2.13 and 2.14, respectively, are shown below:

$$\mu_{eff} = \frac{\mu_{TIL} G_{TIL} + \mu_{SUB} G_{SUB} F^2}{G_{TIL} + G_{SUB} F^2} \tag{D.5}$$

$$n_{eff} = \frac{G_{SH}}{q\mu_{\text{eff}}} = \frac{(G_{TIL} + G_{SUB} F)^2}{q\left(\mu_{TIL} G_{TIL} + \mu_{SUB} G_{SUB} F^2\right)} \tag{D.6}$$

In both equations, μ_{TIL} and μ_{SUB} are the Hall mobilities of the TIL and the substrate, respectively. The Hall mobility and sheet conductance values of the substrate can be experimentally obtained by measuring a substrate without Ti implantation. In this thesis, it is important to remember that the Ti implanted samples included a Back Surface Field (BSF), as explained in the sample preparation section. Thus, the values of G_{TIL} and μ_{TIL} are variables to be fitted as a function of the temperature, applying the bilayer model to the experimental data. Both values are needed to determine the sheet carrier concentration, as described in Eq. 4.9.

In our previous studies, where the KrF excimer laser was used to anneal supersaturated n-type Si samples, the bilayer model accurately fitted most of the experimental data of sheet conductance, Hall mobility and sheet concentration, in a range of temperatures from 7 to 400 K. Several conclusions were drawn, which are briefly described below:

- The decoupling function F usually decreases with temperature, implying that the TIL and the substrate are coupled at high temperatures, but decoupled at low temperatures. The transition temperature in which they start to decouple analytically depends on the temperature dependence of each term of Eq. D.1. Experimentally, it was found to be dependent mainly on the laser fluence used to anneal the sample, as shown by García-Hemme et al. [14], although the Ti dose influenced as well.

- The sheet conductance, sheet concentration and Hall mobility measured at very low temperatures (below 30–40 K) are related to the TIL, as the decoupling function tends to zero in this range for most samples.
- The measured sheet concentration at very low temperatures exhibited values close to the implantation dose, which yielded volumetric concentration values in the order of 10^{20}–10^{21} cm^{-3}, higher than the IB formation limit. Most of the observations led to the conclusion that, in average, each Ti implanted atom contributed with one carrier to the conduction process, with an activation rate close to 100%.
- The measured Hall mobility at temperatures below the transition regime was very low, in the order of 0.1–0.6 $cm^2V^{-1}s^{-1}$, value consistent with a metallic-like layer. Following the band theory, it would imply a high effective mass of carriers, which would be compatible with the idea of a narrow impurity band formed in the TIL.
- The Hall voltage at said range of temperatures changed its sign, indicating that the majority carriers responsible for the conduction process at temperatures lower than the transition temperature were holes instead of electrons (as the substrate used in these studies were n-type).
- The sheet conductance of the TIL did not show a significant dependence with temperature, which could relate to a semi-metallic behaviour of the TIL. In fact, it was found that the sheet conductance value was only slightly higher at room temperature than at very low temperatures.

The main conclusion was that an impurity band might have been formed between the VB and the CB in the TIL, showing a semi-metallic behaviour. It was possible to measure the electrical parameters of the TIL at very low temperatures, where it was determined that the IB of the TIL was contributing with holes to the conduction mechanism.

References

1. Olea J, del Prado A, Pastor D, Martil I, Gonzalez-Diaz G (2011) Sub-bandgap absorption in Ti implanted Si over the Mott limit. J Appl Phys 109:113541
2. Olea J (2009) Procesos de implantación iónica para semiconductores de banda intermedia. Thesis dissertation.
3. Olea J et al (2009) High quality Ti-implanted Si layers above solid solubility limit. In: Proceedings of the 2009 Spanish Conference on Electron Devices, pp 38–41
4. Recht D et al (2013) Supersaturating silicon with transition metals by ion implantation and pulsed laser melting. J Appl Phys 114:124903
5. Liu F, Prucnal S, Hübner R, Yuan Y, Skorupa W, Helm M (2016) Suppressing the cellular breakdown in silicon supersaturated with titanium. J Appl Phys D: Appl Phys 49:5
6. Olea J et al (2016) Room temperature photo-response of titanium supersaturated silicon at energies over the bandgap. J Phys D: Appl Phys 49:055103
7. Garcia-Hemme et al (2014) Room-temperature operation of a titanium supersaturated silicon-based infrared photodetector. Appl Phys Lett 104:211105

8. Chiarotti G, Nannarone S, Pastore R, Chiaradia P (1971) Optical absorption of surface states in Ultrahigh vacuum cleaved (111) surfaces of Ge and Si. Phys Rev B-Solid State 4:3398
9. Olea J et al (2013) Ruling out the impact of defects on the below band gap photoconductivity of Ti supersaturated Si. J Appl Phys 114:053110
10. Maley N (1992) Critical investigation of the infrared-transmission-data analysis of hydrogenated amorphous-silicon alloys. Phys Rev B 46:2078–2085
11. Jellison Jr GE, Lowndes DH (1987) Measurements of the optical properties of liquid silicon and germanium using nanosecond time-resolved ellipsometry. Appl Phys Lett 51:352–354
12. Olea J et al (2011) Two-layer Hall effect model for intermediate band Ti-implanted silicon. J Appl Phys 109:063718
13. Gonzalez-Diaz G et al (2017) A robust method to determine the contact resistance using the van der Pauw set up. Measurement 98:151–158
14. Garcia-Hemme E et al (2015) Meyer Neldel rule application to silicon supersaturated with transition metals. J Phys D: Appl Phys 48:075102

Publications and Scientific Contributions

Scientific Contributions as Peer-Reviewed Manuscripts

1. Pérez E, Castán H, García H, Dueñas S, Bailón L, Montero D, García-Hernansanz R, García-Hemme E, Olea J, González-Díaz G (2015) Energy levels distribution in supersaturated silicon with titanium for photovoltaic applications Appl Phys Lett 106:022105. Quartile Q1. Journal Impact Factor: 3.495
2. Pérez E, Dueñas S, Castán H, García H, Bailón L, Montero D, García-Hernansanz R, García-Hemme E, Olea J, González-Díaz G (2015) A detailed analysis of the energy levels configuration existing in the band gap of supersaturated silicon with titanium for photovoltaic applications. J Appl Phys 118:245704. Quartile Q2. Journal Impact Factor: 2.176
3. García-Hemme E, Montero D, García-Hernansanz R, Olea J, Mártil I, González Díaz G (2016) Insulator-to-metal transition in vanadium supersaturated silicon: variable-range hopping and Kondo effect signatures. J Phys D: Appl Phys 49:275103. Quartile Q2. Journal Impact Factor (2015): 2.772
4. García-Hernansanz R, García-Hemme E, Montero D, Olea J, San Andrés E, del Prado A, Ferrer FJ, Mártil I, González-Díaz G (2016) Limitations of high pressure sputtering for amorphous silicon deposition. Mater Res Express 3(3). Quartile Q2. Journal Impact Factor: 1.151
5. García-Hernansanz R, García-Hemme E, Montero D, del Prado A, Olea J, Andrés E, Mártil I, González-Díaz G (2016) Deposition of intrinsic a-Si:H by ECR-CVD to passivate the crystalline silicon heterointerface in HIT solar cells. IEEE J PhotovoltSl 6(5). Quartile Q1. Journal Impact Factor: 3.736
6. González-Díaz G, Pastor D, García-Hemme E, Montero D, García-Hernansanz R, Olea J, del Prado A, San Andrés E, Mártil I (2017) A robust method to determine the contact resistance using the van der Pauw set up. Measurement 98:151–158. Quartile Q2. Journal Impact Factor: 2.312
7. García-Hemme E, García G, Palacios P, Montero D, García-Hernansanz R, González-Díaz G, Wahnon P (2017) Vanadium supersaturated silicon system: a theoretical and experimental approach. J Phys-D 50(49). Quartile Q2. Journal Impact Factor: 2.373
8. Valdueza-Felip S, Nuñez-Cascajero A, Blasco R, Montero D, de la Mata M, Fernandez S, Rodriguez-De Marcos L, Molina SI, Olea J, Naranjo FB (2018) Influence of the AlN interlayer thickness on the photovoltaic properties of in-rich AlInN on Si heterojunctions deposited by RF sputtering. AIP Adv 8(1):115315. Quartile Q3. Journal Impact Factor: 1.653

D. Montero Álvarez, *Near Infrared Detectors Based on Silicon Supersaturated with Transition Metals*, Springer Theses,
https://doi.org/10.1007/978-3-030-63826-9

9. Olea J, del Prado A, Garcia-Hemme E, Garcia-Hernansanz R, Montero D, Gonzalez-Diaz G, Gonzalo J, Siegel J, Lopez E (2018) Strong subbandgap photoconductivity in GaP implanted with Ti. Prog PhotovoltS 26(3):214–222. Quartile: Q1. Journal Impact Factor: 6.456
10. Garcia H, Castan H, Dueñas S, García-Hemme E, García-Hernansanz R, Montero D, Gonzalez-Diaz G (2018) Energy levels of defects created in silicon supersaturated with transition metals. J Electron Mater 47(9):4993–4997. Quartile: Q3. Journal Impact Factor: 1.566
11. Garcia-Hernansanz R, Garcia-Hemme E, Montero D, Olea J, del Prado A, Martil I, Voz C, Gerling LG, Puigdollers J, Alcubilla R (2018) Transport mechanisms in silicon heterojunction solar cells with molybdenum oxide as a hole transport layer. Sol Energy Mater Sol Cells 185:61–65. Quartile: Q1. Journal Impact Factor: 5.018
12. Blasco R, Nuñez-Cascajero A, Jimenez-Rodriguez M, Montero D, Grenet L, Olea J, Naranjo FB, Valdueza-Felip S (2019) Influence of the AlInN thickness on the photovoltaic characteristics of AlInN on Si solar cells deposited by RF sputtering. Phys Status Solidi A-Appl Mater Sci 216(1):1800494. Quartile Q2. Journal Impact Factor: 1.795
13. Nuñez-Cascajero A, Blasco R, Valdueza-Felip S, Montero D, Olea J, Naranjo FB. High quality $Al_{0.37}In_{0.63}N$ layers grown at low temperature (<300 °C) by radio-frequency sputtering. Naranjo. Mater Sci Semicond Process 100:8–14. Quartile: Q2. Journal Impact Factor: 2.722

Proceedings

1. Garcia-Hernansanz R, Garcia-Hemme E, Montero D, del Prado A, Martil I, Gonzalez-Diaz G, Olea J (2015) Amorphous/crystalline silicon interface characterisation by capacitance and conductance measurements. In: Proceedings of the 2015 IEEE 10th Spanish Conference on Electron Devices (CDE), Aranjuez, Spain

Manuscripts Sent for Peer-Reviewing

1. Olea J, Algaidy S, del Prado A, García-Hemme E, García-Hernansanz R, Montero D, Caudevilla D, González-Díaz G, Soria E, Gonzalo J (2020) On the properties of GaP supersaturated with Ti. J Alloy Compd 820:153358

Contribution to National and International Conferences

1. García-Hernansanz R, García-Hemme E, Olea J, Montero D, del Prado A, Mártil I, González-Díaz G (2015) Amorphous/crystalline silicon interface characterization by capacitance and conductance measurements. In: 10th Spanish Conference on Electron Devices, Aranjuez (Madrid). Oral contribution
2. Perez E, Castán H, Garcia H, Dueñas S, Bailon L, Montero D, García-Hernansanz R, García-Hemme E, Olea J, González-Díaz G (2015) The Meyer-Neldel Rule in the properties of the deep level defects present in silicon supersaturated with titanium. In: 10th Spanish Conference on Electron Devices, Aranjuez (Madrid). Oral contribution
3. Garcia-Hernansanz R, Garcia-Hemme E, Montero D, Olea J, Martil I, Gerling L, Puigdollers J, Alcubilla R, Voz C (2016) Sub-bandgap photoresponse in silicon based solar cell. In: Photovoltaic Technical Conference, Marsella. Oral contribution

4. García-Hemme E, Wang M, Prucnal S, García-Hernansanz R, Montero D, Olea J, González-Díaz G, Zhou S, Skorupa W (2017) Combination of sub-second annealing techniques to reduce lateral segregation of Ti in Si supersaturated material. In: 11th Spanish Conference on Electron Devices, Barcelona. Oral contribution
5. Olea J, García-Hemme E, García-Hernansanz R, Montero D, San-Andrés E and I. Mártil (2017) Hydrogenated amorphous SiTi for photovoltaic applications. 11th Spanish Conference on Electron Devices, Barcelona. Oral contribution
6. García-Hernansanz R, García-Hemme E, Montero D, del Prado A, Mártil I, Voz C, Gerling LG, Puigdollers J, Alcubilla R (2017) Conduction mechanisms in HIT solar cells. In: 11th Spanish Conference on Electron Devices, Barcelona. Oral contribution
7. Montero D, García-Hernansanz R, García-Hemme E, Olea J, San Andrés E, del Prado A, González-Díaz. (2017) Supersaturated silicon based devices for focal plane array detectors for the near and mid IR range at room temperature. In: 11th Spanish Conference on Electron Devices, Barcelona. Oral contribution. Award to best oral communication
8. Blasco R, Nuñez-Cascajero A, Montero D, Olea J, Gonzalez-Herraez M, Naranjo FB, Valdueza-Felip S (2017) Diseño y optimización de células solares de lámina delgada basadas en la heterounión entre el AlInN / silicio depositadas por sputtering. Reunión de Jóvenes Investigadores SINFOTON. Madrid. Poster contribution
9. Valdueza-Felip S, Nuñez-Cascajero A, Blasco R, Montero D, Grenet L, Rodriguez L, Mendez JA, Olea J, Gonzalez-Herraez M, Monroy E, Naranjo FB (2017) Electrical characterization of AlInN on Silicon heterojunctions with AlN buffer layer deposited by sputtering. In: Reunión de Jóvenes Investigadores SINFOTON. Madrid. Poster contribution.
10. Valdueza-Felip S, Nuñez-Cascajero A, Blasco R, Montero D, Grenet L, Rodriguez L, Mendez JA, Olea J, Gonzalez-Herraez M, Monroy E, Naranjo FB (2017) Effect of AlN interfacial layer on the photovoltaic properties of AlInN on silicon heterojunctions deposited by sputtering. In: 10ª Reunión Española de Optoelectrónica (OPTOEL17). Santiago de Compostela. Poster contribution
11. Blasco R, Néñez-Cascajero A, Montero D, Olea J, González-Herráez M, Naranjo FB, Valdueza-Felip S (2017) Design and optimization of thin film AlInN on silicon junctions deposited by sputtering for solar cells. In: 10ª Reunión Española de Optoelectrónica (OPTOEL17). Santiago de Compostel. Poster contribution
12. Valdueza-Felip S, Núñez-Cascajero A, Blasco R, Montero D, Grenet L, Rodríguez L, Méndez JA, Olea J, González-Herráez M, Monroy E, Naranjo FB (2017) Photovoltaic properties of AlInN on silicon heterojunctions deposited by sputtering: effect of AlN interfacial layer. In: 2nd International Conference on Nitride Semiconductors (ICNS 12), Estrasburgo, Francia. Poster contribution
13. García H, Castán H, Dueñas S, García-Hemme E, arcía-Hernansanz R, Montero D, González-DíazG (2017) Energy levels of defects created in silicon supersaturated with transition metals. In: 17th Conference on Defects-recognition, Imaging and Physics in Semiconductors. Valladolid. Poster contribution
14. Blasco R, Núñez-Cascajero A, Montero D, Grenet LJ, Olea J, González-Herráez M, Monroy E, Naranjo FB, Valdueza-Felip S (2018) Influence of the AlInN thickness on the photovoltaic characteristics of AlInN on Si solar cells deposited by RF sputtering. In: Compound Semiconductor Week 2018, Boston, EE. UU. Poster contribution
15. Garcia-Hemme E, Wang M, Berencen Y, Xu C, Garcia-Hernansanz R, Montero D, Algaidy S, Prucnal S, Helm M, Zhou S (2018) Tellurium-hyperdoped silicon for room-temperature short-wavelength infrared detection. In: IEEE 12th Spanish Conference on Electron Devices (CDE), Salamanca, Spain. Oral contribution
16. Garcia-Hernansanz R, Cordero D, Garcia-Hemme E, Montero D, Olea J, del Prado A, San Andres E, Martil I, Voz C, Gerling LG, Puigdollers J, Alcubilla R (2018) Capacitance characterisation of a heterojunction n-type silicon solar cell with MoOx hole-selective contact. In: IEEE 12th Spanish Conference on Electron Devices (CDE), Salamanca, Spain. Oral contribution
17. Montero D, Blasco R, Caudevilla D, Algaidy S, Garcia-Hernansanz R, Garcia-Hemme E, Valdueza-Felip S, Olea J (2018) Influence of Ti concentration on electro-optical properties

of p-type silicon substrates. In: IEEE 12th Spanish Conference on Electron Devices (CDE), Salamanca, Spain. Oral contribution

18. Blasco R, Montero D, Braña A, Olea J, Valdueza-Felip S, Naranjo FB (2019) Effect of Al content on AlxIn1-xN-on-Silicon (x ~ 0–0.6) solar cells deposited by RF Sputtering. In: 13th International Conference on Nitride Semiconductors (ICNS-13). Bellevue, Washington, EE.UU. Poster contribution

Printed in the United States
by Baker & Taylor Publisher Services